New Avenues of Research for Nanoparticle Drug Delivery Systems

New Avenues of Research for Nanoparticle Drug Delivery Systems

Editor

Rafael Prado-Gotor

MDPI • Basel • Beijing • Wuhan • Barcelona • Belgrade • Manchester • Tokyo • Cluj • Tianjin

Editor
Rafael Prado-Gotor
University of Seville
Spain

Editorial Office
MDPI
St. Alban-Anlage 66
4052 Basel, Switzerland

This is a reprint of articles from the Special Issue published online in the open access journal *Nanomaterials* (ISSN 2079-4991) (available at: https://www.mdpi.com/journal/nanomaterials/special_issues/Nanoparticle_Drug_Delivery).

For citation purposes, cite each article independently as indicated on the article page online and as indicated below:

LastName, A.A.; LastName, B.B.; LastName, C.C. Article Title. *Journal Name* **Year**, *Volume Number*, Page Range.

ISBN 978-3-0365-7956-6 (Hbk)
ISBN 978-3-0365-7957-3 (PDF)

Contents

About the Editor

Rafael Prado-Gotor

R. Prado-Gotor received a B.S. in Chemistry from the University of Seville. In 1999, he received his Ph.D. from the same university. He carried out postdoctoral research for two years (2000–2001) under Prof. M. Kochoyan and C. Roumestand at the Centre of Structural Biochemistry (C.N.R.S., Montpellier) on the development of the use of NMR to study DNA and proteins in solution. Since 2000, he has been a Research Professor at the department of Physical Chemistry in the University of Seville. In 2003, he worked with Prof. Soledad Penades in the new Research Centre of La Isla de la Cartuja (CICIC). From this moment on, his research has focused on the study of the interactions of proteins and DNA with nanoparticles and on the optimization of these interactions from thermodynamic and kinetic points of view. He has over one hundred publications, including six book chapters and three patents in the field of nanoparticles with biomedical applications, and more than sixty participations in congresses, being the PI responsible for the FQM 386 Group "Synthesis and Physicochemistry of Nanoparticles".

nanomaterials

Editorial

New Avenues of Research for Nanoparticle Drug Delivery Systems

Rafael Prado-Gotor

Department of Physical Chemistry, Faculty of Chemistry, University of Seville, C/Profesor García González 1, 41012 Seville, Spain; pradogotor@us.es

Knowledge of the different elements that determine the optimal method for drug loading and delivery nanosystems using nanoparticles of different natures is experiencing a remarkable boom in many scientific fields, especially in medicine, chemistry, biology, materials science and molecular biotechnology. Not only is it necessary to fully understand what type of nanoparticle is most appropriate in each particular situation, but the interactions of the used nanoparticle with the system under study or the analysis of the release process are important as well. To design nanoparticle drug delivery agents well, a high ratio of drug to nanoparticles must be achieved. Furthermore, biocompatibility and knowledge of the time the release process takes are fundamental factors for the construction of operative NPs as drug delivery vectors.

This Special Issue, "New Avenues of Research for Nanoparticle Drug Delivery Systems", brings together recent research articles published in *Nanomaterials*. Specifically, five original research articles, one of them a review, were published by authors from different countries on what is a current issue in this field of research.

S. E. Kim et al. [1] prepared and characterized lactoferrin (LF) conjugated carboxylated-nanodiamonds (cNDs) in order to analyse their possible effects against oxidative stress, inflammatory response and osteogenic differentiation of cells. LF-NDs not only markedly suppress reactive forms of oxygen (ROS) in cells, but they also shelter cells in ROS environments and can significantly decrease the levels of pro-inflammatory cytokines (IL-1β and TNF-α) secreted by cells challenged with LPS. Furthermore, LF-NDs promote osteogenic differentiation of MC3T3-E1 cells by increasing alkaline phosphatase activity and calcium deposition via LF release.

L. Gómez-Segura et al. [2] investigated the ex vivo permeation of carprofen (CP) 2-(6-chlorocarbazole) propionic acid across different classes of porcine mucous membranes (buccal, sublingual and vaginal) and ophthalmic linings (cornea, sclera and conjunctiva) to differentiate between the CP-NP formulation and a CP solution (CP-Solution). The results showed that CP-NPs provide advantageous situations in most tissues. The structure of the material is not modified, being more effective and safer than the CP solution. This research opens the door to successfully test in situ treatments of many inflammatory diseases in animals or people.

P. M. Castillo et al. [3], working in the field of disease detection using nanosystems, have carried out a new method to determine urinary lysozyme content within a concentration range that is associated with monocytic and myelomonocytic leukemia, among other diseases. The authors describe a method based on obtaining CIELab parameters described by the CIE. The advantages of the method involve a very low cost and an extremely short detection time. In addition to being fast (less than 10 min) and economical, the described method does not require specialized personnel in the knowledge of specific analytical techniques. The required equipment implies having a commercial spectrophotometer or colorimeter, and a positive case can even be detected through an evaluation of the color with the naked eye with a reference solution.

Citation: Prado-Gotor, R. New Avenues of Research for Nanoparticle Drug Delivery Systems. *Nanomaterials* **2022**, *12*, 4141. https://doi.org/10.3390/nano12234141

Received: 26 October 2022
Accepted: 18 November 2022
Published: 23 November 2022

A. Gomes et al. [4] have developed an exhaustive study of the interaction energy between lysozyme and the surface of gold nanoparticles surrounded by citrate ions, considering with special attention those factors that can directly influence the performance of colloidal gold systems for the detection of the protein. Specifically, authors have analysed the stability of colloidal gold solutions, the influence of the diameter of the nanoparticles and the correct way to express the concentration of gold in the nanosystem to obtain the best and most accurate results. They found that the state of saturation implies an average number of 55 Lys per gold nanoparticle. On the other hand, it has been found that the free energy (ΔG^0) corresponding to the interaction of the protein with the 10 nm gold surface is about -40.8 kJ mol^{-1}.

B. Begines et al. [5] have prepared a review based on polymeric nanoparticles for drug delivery. This review focuses on the most recent applications and advances of nanoparticulate polymeric formulations as nanocarriers, especially those used to battle specific diseases. Studies about the use of different materials as nanocarriers should meet important requisites such as biocompatibility, biodegradability and non-immunogenicity. The toxicity associated with numerous drugs and classical galenic formulations or the complexity to treat diseases have progressed the prompt development of new alternatives to drug-eluting nanosystems. In this sense, polymers are macromolecules synthetized via a covalent union of one or different monomers that possess at least two functional groups, where they can react easily to constitute a chain to attain specific properties. Polymeric nanoparticles are not only pharmaceutical entities that may exhibit all the above-mentioned characteristics: their rich synthetic versatility allows them to be greatly customized to accomplish the final requirements. Particularly, polymeric nanoparticles for ocular drug delivery, for cancer diagnosis and treatment, as well as nutraceutical delivery, have been described in detail, as well as an interesting discussion concerning the future prospects of these systems.

I would like to thank all the authors and reviewers of this Special Issue. I also acknowledge the Assistant Editor, Riven Yang, for her trust, support and effort in moving this issue forward. In addition, authors are encouraged to submit original research articles and reviews in the next Special Issue: "Recent Advances in Targeted Therapy Using Multifunctionalized Gold Nanoparticles".

Funding: This work was financed by the Consejería de Innovación, Ciencia y Empresas de la Junta de Andalucía (2021/FQM-386) and by the VII PP USO SSGG ((2020/00001068), (2020/00001073) and (2022/00000274)) of the University of Seville.

Data Availability Statement: Not applicable.

Conflicts of Interest: The author declares no conflict of interest.

References

1. Kim, S.E.; Choi, S.; Hong, J.-Y.; Shim, K.-S.; Kim, T.-H.; Park, K.; Lee, S.-H. Accelerated Osteogenic Differentiation of MC3T3-E1 Cells by Lactoferrin-Conjugated Nanodiamonds through Enhanced Anti-Oxidant and Anti-Inflammatory Effects. *Nanomaterials* **2020**, *10*, 50. [CrossRef]
2. Gómez-Segura, L.; Parra, A.; Calpena-Campmany, A.C.; Gimeno, Á.; Gómez de Aranda, I.; Boix-Montañes, A. Ex Vivo Permeation of Carprofen Vehiculated by PLGA Nanoparticles through Porcine Mucous Membranes and Ophthalmic Tissues. *Nanomaterials* **2020**, *10*, 355. [CrossRef] [PubMed]
3. Castillo, P.M.; Fernández-Acejo, F.J.; Carnerero, J.M.; Prado-Gotor, R.; Jimenez-Ruiz, A. Colorimetric, Naked-Eye Detection of Lysozyme in Human Urine with Gold Nanoparticles. *Nanomaterials* **2021**, *11*, 612. [CrossRef] [PubMed]
4. Gomes, A.; Carnerero, J.M.; Jimenez-Ruiz, A.; Grueso, E.; Giráldez-Pérez, R.M.; Prado-Gotor, R. Lysozyme–AuNPs Interactions: Determination of Binding Free Energy. *Nanomaterials* **2021**, *11*, 2139. [CrossRef] [PubMed]
5. Begines, B.; Ortiz, T.; Pérez-Aranda, M.; Martínez, G.; Merinero, M.; Argüelles-Arias, F.; Alcudia, A. Polymeric Nanoparticles for Drug Delivery: Recent Developments and Future Prospects. *Nanomaterials* **2020**, *10*, 1403. [CrossRef] [PubMed]

nanomaterials

MDPI

Article

Accelerated Osteogenic Differentiation of MC3T3-E1 Cells by Lactoferrin-Conjugated Nanodiamonds through Enhanced Anti-Oxidant and Anti-Inflammatory Effects

Sung Eun Kim [1,†], Somang Choi [1,2,†], Jae-Young Hong [3,†], Kyu-Sik Shim [2], Tae-Hoon Kim [4], Kyeongsoon Park [5] and Suk-Ha Lee [4,*]

1 Department of Orthopedic Surgery and Rare Diseases Institute, Korea University Guro Hospital, #148, Gurodong-ro, Guro-gu, Seoul 08308, Korea; sekim10@korea.ac.kr (S.E.K.); chlthakd1029@naver.com (S.C.)

2 Department of Biomedical Science, College of Medicine, Korea University, Anam-dong, Seongbuk-gu, Seoul 02841, Korea; breakdown88@korea.ac.kr

3 Department of Orthopedic Surgery, Korea University Ansan Hospital, #123, Jeokgeum-ro, Danwon-gu, Ansan-si, Gyeonggi-do 15355, Korea; osspine@korea.ac.kr

4 Department of Orthopedic Surgery, Konkuk University Medical Center, Konkuk University School of Medicine, #120-1, Neungdong-ro, Hwayang-dong, Gwangjin-gu, Seoul 05030, Korea; thkim96@hanmail.net

5 Department of Systems Biotechnology, College of Biotechnology and Natural Resources, Chung-Ang University, 4726 Seodong-daero, Daedeok-myeon, Anseong-si, Gyeonggi-do 17546, Korea; kspark1223@cau.ac.kr

* Correspondence: 24926@kuh.ac.kr; Tel.: +82-2-2030-7611

† These authore contributed equally to this work.

Received: 2 December 2019; Accepted: 18 December 2019; Published: 24 December 2019

Abstract: The purpose of this study was to investigate the effects of lactoferrin (LF)-conjugated nanodiamonds (NDs) in vitro on both anti-oxidant and anti-inflammation activity as well as osteogenic promotion. The application of LF-NDs resulted in sustained release of LF for up to 7 days. In vitro anti-oxidant analyses performed using Dichlorofluorescin diacetate (DCF-DA) assay and cell proliferation studies showed that LF (50 μg)-NDs effectively scavenged the reactive oxygen species (ROS) in MC3T3-E1 cells (osteoblast-like cells) after H_2O_2 treatment and increased proliferation of cells after H_2O_2 treatment. Treatment of lipopolysaccharide (LPS)-induced MC3T3-E1 cells with LF-NDs suppressed levels of pro-inflammatory cytokines, including interleukin-1β (IL-1β) and tumor necrosis factor-α (TNF-α). In addition, LF-NDs were associated with outstanding enhancement of osteogenic activity of MC3T3-E1 cells due to increased alkaline phosphatase (ALP) and calcium deposition. Our findings suggest that LF-NDs are an important substrate for alleviating ROS effects and inflammation, as well as promoting osteogenic differentiation of cells.

Keywords: nanodiamonds; lactoferrin; anti-oxidant; anti-inflammatory; osteogenic differentiation

1. Introduction

Bone, plays major roles in the support, movement and protection of bodily organs and is a dynamic tissue with renewal and repair. Despite these properties, bone defects that result from trauma, traffic accidents (TA), congenital deformation, disease, and fracture sometimes require bone grafts. Autografts are considered the gold standard of bone graft replacements, due to their osteoinductive, osteoconductive and osteogenic properties. However, they also have shortcomings such as restricted supply, donor-site morbidity and accompanying pain [1,2]. Other options for treating bone defects are allografts, which are characterized by osteoinductive and osteoconductive characteristics. Major drawbacks of allografts are associated with blood loss, disease transmission, and cost [3,4].

The activities of osteoblasts and osteoclasts control bone remodeling via bone formation and resorption. However, there have been reports that inhibition of osteoblast differentiation and induction of apoptosis may occur due to oxidative stressors such as reactive oxygen species (ROS) [5–7]. In addition, the elevated levels of ROS in osteoblasts are associated with inhibition of mineralization and delayed bone healing [8–10].

Lactoferrin (LF), an iron-binding glycoprotein that is isolated from human and bovine milks, belongs to the transferrin family [11]. Oral administration of LF influences iron absorption and metabolism [12]. LF production increases in an inflammatory circumstance, which in turn indicates that LF has anti-inflammatory activity. LF is known as an antioxidant protein that increases antioxidant capacity and reduces ROS formation [13–15]. LF has pleiotropic effects including immunomodulatory, anticancer, antibacterial and antiviral activities [16–18]. Interestingly, LF is known to have potential as an osteogenic factor. Previous reports have demonstrated that materials functionalized with LF induce osteogenic differentiation of mesenchymal stem cells (MSCs) and adipose-derived stem cells (ADSCs) [19,20]. The mechanism of LF action on osteogenic differentiation stimulates cells via lipoprotein receptor-related protein 1 (LRP1)-independent protein kinase A (PKA) and p38 signaling pathways [21]. Therefore, to improve its therapeutic efficacy, LF should be incorporated or immobilized within/on substrates.

During the last 10 years, nanomaterials have been utilized in biomedical applications [22,23]. Among these, carbon-based nanomaterials such as carbon nanotubes, fullerenes, graphene and nanodiamond are major components of all living things [24]. Nanodiamond (ND) is considered an excellent material due to its good biocompatibility, low toxicity, hardness, and high surface functionality [24,25]. Moreover, ND can be facilitated to anchor drugs or biomolecules in different environments for delivery and sustained release due to non-covalent or covalent interactions [26,27]. Multifunctional ND can be used as a bone scaffold in combination with different polymers. Bone scaffolds fabricated by Zhang et al. [27] using PLLA polymer and octadecylamine-functionalized NDs (ODA-NDs) showed no harmful results on cell proliferation when murine osteoblasts were cultured on bone scaffolds for more than 1 week. Following this investigation, such scaffolds led to enhancement of biomechanical properties, such as a 280% increase in failure strain and a 310% improvement of the fracture force in tensile strength [28]. In other studies performed by Parizek et al. [29] and Ahn et al. [30], ND-composited PLGA nanofibrous membranes and PCL fibrous matrices not only enhanced mechanical resistance but also induced proliferation and differentiation of osteoblast-like cells.

In the present study, we designed NDs-based nanoparticles with anti-oxidant, anti-inflammatory and osteogenic effects by anchoring LF. Here, we demonstrate the utility of multi-functional ND with LF by anti-oxidant, IL-1β and TNF-α assays, alkaline phosphatase (ALP) activity, and calcium content against MC3T3-E1 cells.

2. Materials and Methods

2.1. Lactoferrin (LF) Conjugated Carboxylated-Nanodiamonds (cNDs)

For the conjugation of human lactoferrin (LF, Sigma-Aldrich, St. Louis, MO, USA), 10 mg of carboxylated-nanodiamonds (cNDs, Tokyo Chemical Industry Co., Ltd., Tokyo, Japan) were placed in a sterile PBS solution (pH 7.4) and gently stirred for 30 min at room temperature (RT), followed by the addition of LF (10 or 50 $\mu g \cdot mL^{-1}$) and incubated for 24 h. After incubation for 24 h, the product was rinsed three times with PBS solution and centrifuged at 3000 rpm for 10 min at 4 °C using Micro Refrigerated Centrifuge (Smart R17, Hanil Science Industrial, Incheon, Korea). The supernatant was collected to analyze the loading amount of LF on cND surface and the sample was lyophilized using freeze dryer (FD8508, IlShinBioBase Co., Ltd. Gyeonggido, Korea) for 3 days. To quantify the loading amount of LF, the collected supernatant after the LF conjugated cNDs was measured with a Pierce bicinchoninic acid (BCA) protein assay kit (Thermo Fisher Scientific, Rockford, IL, USA) following the manufacturer's protocol. In brief, the supernatant (25 μL) was placed in e-tube containing 200 μL of BCA reagent and incubated at RT for 1 h at 37 °C. After incubation, 100 μL of solution was transferred to 96-well plates and monitored at 562 nm using a Flash Multimode Reader (Varioskan^{TM}, Thermo Scientific, Waltham, MA, USA). The loading amount of LF on the ND surface was evaluated by determining the loading amount of LF relative to the initial LF amount. cNDs, LF (10 μg) conjugated cNDs, and LF (50 μg) conjugated cNDs were designated as NDs, LF (10 μg)-NDs and LF (50 μg)-NDs, respectively.

2.2. Characterization

Morphologies of NDs with or without conjugated LF were visualized with a transmission electron microscope (TEM, JEM-F200, JEOL Ltd., Tokyo, Japan) at the Yonsei Center for Research Facilities. Prior to the observation of TEM, each sample was pre-treated as follows: 10 μg of each sample was dispersed in an e-tube containing 1 mL ethanol (EtOH), followed by sonication using Powersonic 405 (bath-type instruments; 40 KHz, power: 350 W, Hwashin Tech Co., Ltd., Seoul, Korea) for 1 h at 4 °C. Each sample was then pipetted onto a copper TEM grid (CF200-Cu, Electron Microscopy Sciences, Hatfield, PA, USA) and the solvent of each sample was removed after overnight deposition. TEM was accelerated at 200 kV.

To confirm the size distribution of each sample, 100 $\mu g \cdot mL^{-1}$ of each sample was suspended using deionized and distilled water (DDW) and sonicated using a Powersonic 405 for 1 h at 4 °C. Then, 1 mL of each dispersed sample was pipetted onto cuvettes (CUVEETTA STD UV 4 FACCE OTT. Kartell S.p.A., Noviglio, Italy). Dynamic light scattering (DLS) analysis was conducted to assess the size distribution of each sample using a Malvern Matersizer 3000 instrument (Malvern Panalytical Ltd., Malvern, UK) with a He-Ne laser at a wavelength of 633 nm. For zeta potential measurements, 1 mL of each dispersed sample was placed in a disposable micro-cuvette (Zetasizer Nano Series, Malvern Panalytical Ltd.). The zeta potential measurements were also performed on Malvern Matersizer 3000 instrument. In order to observe the surface chemical compositions of NDs with or without LF, we performed X-ray photoelectron spectroscopy (XPS) with a K-alpha spectrometer (ESCALAB250 XPS System, Theta Probe AR-XPS System, Thermo Fisher Scientific, Waltham, MA, USA) with 1486.6 eV photons as an Al Kα X-ray source at the Korea Basic Science Institute Busan Center. The surface compositions of different test groups were analyzed using attenuated total reflectance Fourier transform infrared (ATR-FTIR, Avatar 360, Nicolet Instrument Corp., Madison, WI, USA) spectroscopy with a resolution of 4 cm^{-1} between 4000 and 600 cm^{-1}. The ATR-FTIR spectrum of LF alone was observed to measure a commercially available powder.

2.3. In Vitro LF Release

In order to assess the release of LF from LF (10 μg)-NDs and LF (50 μg)-NDs, we performed assays at pre-designated time intervals. At each interval, 10 mg of each sample was dispersed directly in 1 mL of PBS solution (pH 7.4) and pipetted in a dialysis bag (cutoff molecular weight 6000–8000 Da). The bag was placed in a conical tube containing 5 mL PBS, followed by shaking at a speed of 100 rpm at 37 °C. PBS solution used as a release medium was harvested at pre-designated time intervals and replaced by the same volume of fresh PBS. The amount of LF released was monitored with a Pierce BCA protein assay kit using a Flash Multimode Reader at 562 nm.

2.4. Cytotoxicity

Before the determination of the effect of LF-NDs on the cellular activity, we performed cytotoxicity tests of all samples. Briefly, MC3T3-E1 cells (Korean Cell Line Bank, Seoul, Korea) were seeded at a density of 1×10^4 cells per well in 96-well plates and cultured with Dulbecco's modified Eagle's medium (DMEM, Thermo Fisher Scientific Inc., USA) containing 10% fetal bovine serum (FBS, Thermo Fisher Scientific Inc., USA) and 1% antibiotics (100 $U \cdot mL^{-1}$ penicillin and 0.1 $mg \cdot mL^{-1}$ streptomycin, Thermo Fisher Scientific Inc., USA) in 5% CO_2 at 37 °C. After culturing for 24 h, cells were washed with PBS and exposed to each sample (100 $\mu g \cdot mL^{-1}$). After cultivation for 24 and 48 h, DMEM in cells was aspirated and PBS buffer was added to wash cells. Then, 10 μL of 1-(4,5-dimethylthiazol-2-yl)-3,5-diphenyl formazan (MTT) reagent (Sigma-Aldrich) was added to each well and incubated for 4 h in the dark at 37 °C. At the end of the incubation period, 150 μL of dimethyl sulfoxide (DMSO, Sigma-Aldrich, St. Louis, MO, USA) was added to each well to dissolve the formazan crystals and absorbance was read at 595 nm with a Flash Multimode Reader. Medium from cells without samples were used. The cytotoxicity was represented as the percentage of viable cells vs. the control.

2.5. *Cellular Uptake Anaalysis*

In order to assess the cellular uptake of cNDs, cNDs were conjugated with fluorescein isothiocyanate (FITC, Thermo Fisher Scientific, USA). Before FITC conjugation, the cND surface was first modified by dopamine (Sigma-Aldrich) to anchor the amine group. Briefly, cNDs (10 mg) were suspended in 10 mL of 10 mM Tris·HCl (pH 8.0), dissolving 10 mg of dopamine, and the mixture was gently stirred while avoiding light exposure for 24 h. After reaction, NH_2-NDs were washed several times with DDW and lyophilized for 3 days. Ten mg of NH_2-NDs were dispersed in MES buffer (pH 4.5) and 100 μg of FITC was added to the buffer. Mixtures were maintained at RT in the dark overnight. The products were centrifuged, and the sediments were rinsed with DDW and lyophilized for 2 days. Cells at a density of 1×10^4 cells per dish were seeded on microscope cover glasses (12 mm diameter, Paul Marienfeld GebH & Co., Lauda-Königshofen, Germany) and adhered for 24 h. After adhesion for 24 h, cells were rinsed twice with PBS and exposed to FITC-conjugated NDs (100 $\mu g \cdot mL^{-1}$), followed by incubation at 37 °C and 5% CO_2 for 4 h. After incubation, cells were fixed with 4% paraformaldehyde for 30 min. Rhodamine-phalloidin (1:200, Thermo Fisher Scientific, USA) and 4-6-diamidino-2-phenylindole (DAPI, Thermo Fisher Scientific, USA) were added to cells for staining cell nuclei for 30 min at RT. Then, the samples were observed by a confocal laser scanning microscope (CLSM, LSM700, Zeiss, Germany).

2.6. *Evaluation of ROS Scavenging Activity*

2.6.1. Suppression of ROS at the Cell Level

In order to measure the ROS scavenging capacity of LF-NDs in cells, we performed 2′,7-dichlorodihydrofluorescein diacetate (DCFDA) staining and DCFDA assays. MC3T3-E1 was seeded at a concentration of 1×10^4 cells per well at microscope cover glasses in 24-well plates and incubated for 24 h. After incubation for 24 h, cells were treated with 300 μM H_2O_2 at 37 °C for 30 min, followed by the aspiration of 300 μM H_2O_2. Cells were treated using the extracted DMEM without FBS from each sample for 24 h. At pre-designed times of 6 or 24 h, the cells were stained with DCFDA (25 μM) for 45 min in the dark, rinsed with PBS, and fixed with 3.7% paraformaldehyde for 20 min. Cell images were observed using a confocal laser scanning microscope. In order to further quantify the ROS levels in cells served with the extracted DMEM from each group, cells were examined with a DCFDA/H2DCFDA cellular ROS assay kit (Abcam, Cambridge, MA, USA) in accordance with the manufacturer's protocols. The quantitative fluorescence spectra were recorded by a Flash Multimode Reader with excitation/emission at 495 nm/529 nm.

2.6.2. Protection of Cell Suppression in the ROS Condition

Cell survival capacity of MC3T3-E1 cells treated with each sample in ROS condition was analyzed using MTT reagent. The cells were seeded in 24-well plates at a density of 1×10^5 cells per well and incubated with DMEM in the presence or absence of each sample (100 $\mu g \cdot mL^{-1}$) for 24 h. After 24 h, the cells were exposed to 300 μM H_2O_2 at 37 °C for 30 min, followed by another incubation for 6 or 24 h. The cells were treated with MTT reagent for 4 h at 37 °C in 5% CO_2. After 4 h incubation, formazan crystals were formed, followed by the addition of DMSO to dissolve the formazan crystals. The solution was added to 96-well plates and monitored at 595 nm with a Flash Multimode Reader.

2.7. Interleukin-1β (IL-1β) and Tumor Necrosis Factor Alpha (TNF-α) Content

In order to assess the anti-inflammatory activities of LF-NDs, 1×10^5 cells were seeded in each well of a 24-well plate with DMEM and treated with 100 $\mu g \cdot mL^{-1}$ of LPS in the presence or absence of each sample (100 $\mu g \cdot mL^{-1}$). At pre-designed time points, the supernatants were harvested and stored at −20 °C for further quantification of IL-1β and TNF-α. The amount of IL-1β and TNF-α secreted in cells was analyzed using enzyme-linked immunosorbent assay (ELISA) kits (BioGems Ltd., Westlake Village, CA, USA). The absorbance value was monitored at 450 nm using a Flash Multimode Reader.

2.8. Alkaline Phosphatase (ALP) Activity

MC3T3-E1 cells (1×10^5 $cells \cdot mL^{-1}$) were seeded on 24-well plates and exposed to each sample at a concentration of 100 $\mu g \cdot mL^{-1}$. At the end of each period, cells were lysed with lysis buffer (1× RIPA buffer) and transferred to e-tubes. Cell lysates were centrifuged at 13,500 rpm for 10 min at 4 °C using a Micro Refrigerated Centrifuge (Smart R17, Hanil Science Industrial, Incheon, Korea). Supernatant was transferred to new e-tubes and P-nitrophenyl phosphate (Sigma-Aldrich, USA) solution was added, followed by incubation at 37 °C for 30 min. After incubation for 30 min, 500 μL of 1N NaOH was added to the solution to stop the reaction. The absorbance was evaluated at 405 nm with a Flash Multimode Reader. Total protein concentration was normalized using Bradford reagent (Bio-Rad Laboratories, Inc., Hercules, CA, USA) and bovine serum albumin (BSA, Bio-Rad Laboratories, Inc.).

2.9. Calcium Deposition

MC3T3-E1 cells (1×10^5 $cells \cdot mL^{-1}$) were seeded in 24-well plates and cultured with each sample (100 $\mu g \cdot mL^{-1}$). After being exposed to each sample at pre-determined time intervals, cells were rinsed three times with PBS and 500 μL of 0.5N·HCl was added to cells, followed by incubation at 100 rpm overnight at 37 °C using shaking incubator (SI-300R, Jeio Tech Co., Ltd., Seoul, Korea). After overnight incubation, each sample (20 μL) was transferred to an e-tube and calcium standard solution (20 μL) was added to the sample solution, following the addition of color reagent solution (400 μL) [25 mg of o-cresolphthalein complexone (Sigma-Aldrich) and 250 mg of 8-hydroxy-quindine (Sigma-Aldrich)]. The resulting solution was vortexed for 1 min and then 20 μL of AMP buffer [37.8 mL of 2-amino-2-methyl-1-propanol (Sigma-Aldrich)] was added and reacted for 15 min at RT. After the reaction, the solution (200 μL) was carefully transferred to 96-well plates and absorbance was recorded with a Flash Multimode Reader at 575 nm.

2.10. Statistical Analysis

Data are presented as mean ± standard deviation. Statistical comparisons were performed via one-way analysis of variance (ANOVA) using Systat software (Chicago, IL, USA). Differences were considered statistically significant at * $p < 0.05$ and ** $p < 0.01$.

3. Results

3.1. Characterization of NDs with and without LF

The TEM images in Figure 1 distinctly demonstrate the morphologies of NDs with and without LF. Each ND group was round shape and nano-sized. The diameters and size distributions of NDs, LF (10 μg)-NDs, and LF (50 μg)-NDs were investigated by DLS.

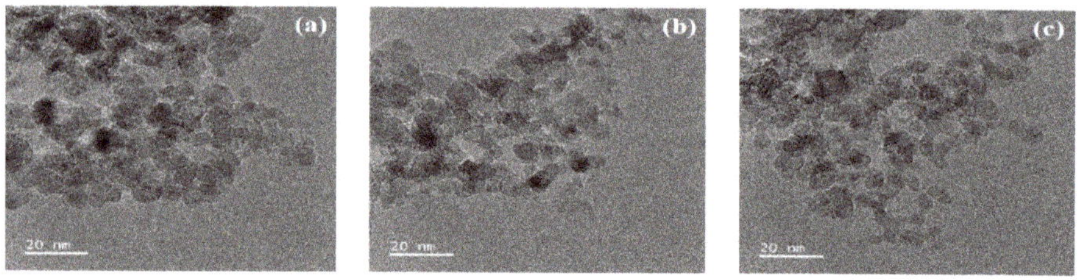

Figure 1. Transmission electron microscope (TEM) images of (**a**) NDs; (**b**) LF (10 μg)-NDs; (**c**) LF (50 μg)-NDs. Scale bar: 20 nm.

As shown in Figure 2A, the particle sizes and distributions were 209.00 ± 103.10 nm with a polydispersity index (PDI) of 0.203 for NDs, 211.30 ± 95.22 nm with PDI of 0.189 for LF (10 μg)-NDs, and 216.50 ± 108.50 nm with PDI of 0.183 for LF (50 μg)-NDs, respectively. Smaller hydrodynamic diameter and narrower PDI of LF-NDs were observed, which suggests that LF-NDs have improved dispersibility due to protein conjugating. The zeta potential values of NDs, LF (10 μg)-NDs, and LF (50 μg)-NDs were −26.23 ± 0.80, −27.77 ± 1.30 and −28.50 ± 0.85 mV, respectively.

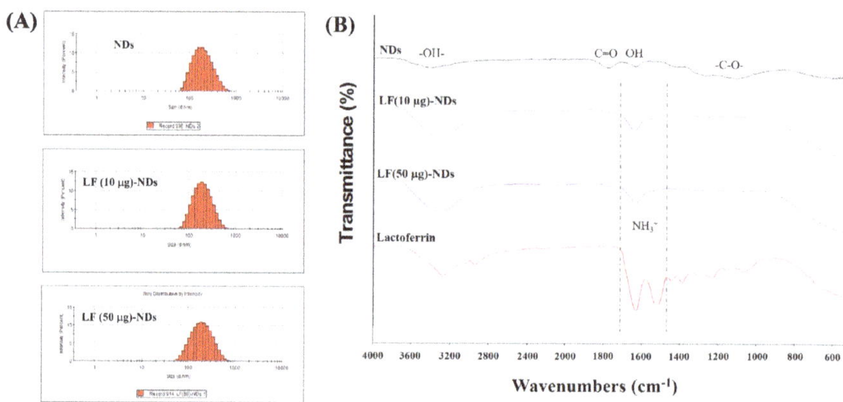

Figure 2. (**A**) Particle size and distribution of NDs, LF (10 μg)-NDs and LF (50 μg)-NDs measured by dynamic light scattering (DLS); (**B**) Attenuated total reflectance Fourier transform infrared (ATR-FTIR) spectra of NDs, LF (10 μg)-NDs, LF (50 μg)-NDs and Lactoferrin.

XPS was conducted to confirm the surface chemical compositions of NDs, LF (10 μg)-NDs and LF (50 μg)-NDs (Table 1). NDs conjugated by LF (10 or 50 μg) were confirmed by increases in N1s component from 1.83% to 3.09% and 1.83% to 4.74, respectively, indicating that LF is existed on the surfaces of the NDs. To further confirm the LF immobilization on the NDs, ATR-FTIR spectra of each group are shown in Figure 2B before and after LF (10 or 50 μg) conjugation. After conjugating with LF, we observed strong absorption new peak bands at 1635 and 1517 cm^{-1}, which correspond to the C=O stretching vibration of amide I and N-H bending vibration of amide II, respectively, suggesting the successful conjugation of LF. The loading amount and efficiency of LF from LF (10 μg)-NDs and LF (50 μg)-NDs were 6.44 ± 0.37 μg (64.39 ± 3.66%) and 41.15 ± 1.94 μg (82.30 ± 3.88%), respectively.

Table 1. Surface elemental composition of NDs, LF (10 μg)-NDs, and LF (50 μg)-NDs.

Sample \ Elements	C1s (%)	N1s (%)	O1s (%)	Total (%)
NDs	87.32	1.83	10.85	100
LF (10 μg)-NDs	85.99	3.09	10.92	100
LF (10 μg)-NDs	84.45	4.74	10.81	100

3.2. In Vitro LF Release

As shown in Figure 3, the in vitro release profiles of LF from LF (10 μg)-NDs and LF (50 μg)-NDs showed sustained release patterns. At 1 day, the released amounts and percentages of LF were 3.60 ± 0.08 μg (55.95 ± 0.57%) for LF (10 μg)-NDs and 25.06 ± 0.42 μg (60.90 ± 1.03%) for (50 μg)-NDs. For the 7-day period, LF (10 μg)-NDs and LF (50 μg)-NDs released 4.54 ± 0.29 μg (70.56 ± 4.44%) and 29.22 ± 0.41 μg (71.00 ± 0.99%) of LF, respectively.

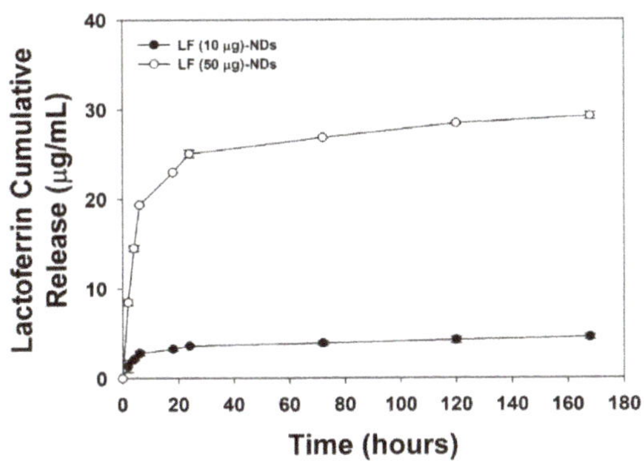

Figure 3. In vitro release profiles of LF from LF (10 μg)-NDs and LF (50 μg)-NDs for 7 days. ($n = 4$).

3.3. Cytotoxicity and Cellular Internalization

Figure 4A shows the cytotoxicity test results for each sample compared against MC3T3-E1 cells at 24 and 48 h. Viabilities of cells treated with each sample were preserved over 98% for 48 h compared to the control group, suggesting that there were no cytotoxic effects on MC3T3-E1 cells in any sample. CLSM was used to confirm the intracellular uptake of NDs with or without LF. Previous study showed that ND particles can be internalized through the cell membrane and accumulate in the cytoplasm [31]. Consistent with the previous results, after 4 h incubation, FITC-conjugated NDs were observed around the cytoplasm and nuclei of cells (Figure 4B).

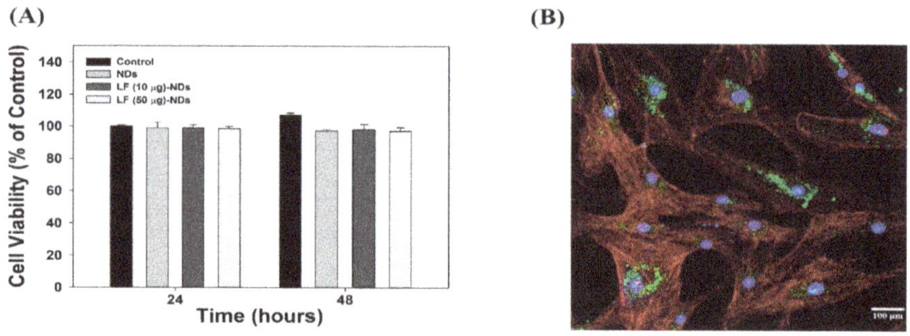

Figure 4. (**A**) Cytotoxicity test of NDs, LF (10 μg)-NDs and LF (50 μg)-NDs against MC3T3-E1 cells for 24 and 48 h; (**B**) In vitro cellular internalization of FITC-NDs after incubation of 4 h. Scale bar: 100 μm.

3.4. ROS Scavenging Effects of LF-NDs in Cells

In order to investigate the anti-oxidant activities of each sample, MC3T3-E1 cells were pre-treated with 300 μM H_2O_2 exposure for 30 min in order to create ROS. Under 300 μM H_2O_2 condition, controls without sample treatment showed high fluorescence intensity in images taken at 6 and 24 h (Figure 5A,B). However, cells treated with extracts of NDs with or without LF showed low fluorescence intensities and images in a time-dependent manner. Treatment with extract from LF (50 μg)-NDs led to the lowest fluorescence intensity and images among groups. These results indicate that LF (50 μg)-NDs have excellent anti-oxidant activity.

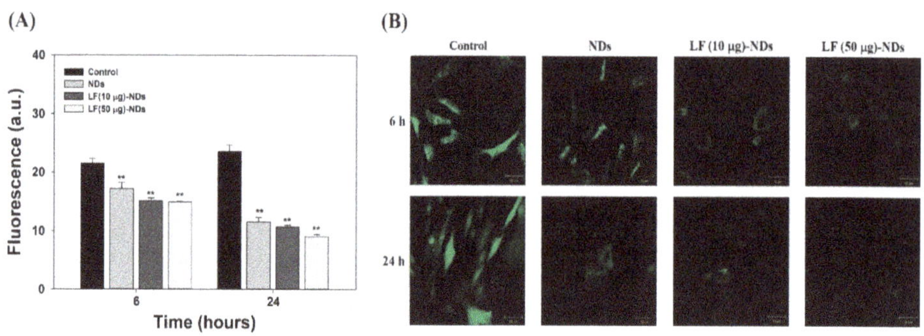

Figure 5. (**A**) Quantitative intracellular ROS levels of MC3T3-E1 cells treated with extract from NDs, LF (10 μg)-NDs and LF (50 μg)-NDs for 6 and 24 h after the cells were treated with 300 μM H_2O_2 for 30 min. The error bars represent mean ± standard deviation ($n = 4$). p value is a comparison between LF-conjugated NDs and NDs. ** $p < 0.01$; (**B**) Fluorescence images of intracellular levels of MC3T3-E1 cells treated with extract from NDs, LF (10 μg)-NDs and LF (50 μg)-NDs for 6 and 24 h after the cells were treated with 300 μM H_2O_2 for 30 min. After 6 and 24 h treatment, the cells were stained with 2′,7-dichlorodihydrofluorescein diacetate (DCFDA) and observed by a confocal laser scanning microscope (CLSM). Scale bar = 50 μm.

3.5. Cellular Protection Against ROS

In order to further demonstrate the direct anti-oxidant effects of LF-NDs in cells, we measured the proliferation of MC3T3-E1 cells treated with 300 μM H_2O_2 in the presence or absence of each sample at 6 and 24 h. As shown in Figure 6, there were significant differences in cell proliferation between MC3T3-E1 cells treated with NDs with or without LF and those of controls at 6 and 24 h, while the cell viability of the control group was reduced to 24 h rather than 6 h. However, viabilities of cells treated with LF-NDs were greater than those treated with NDs in a concentration- and time-dependent manner.

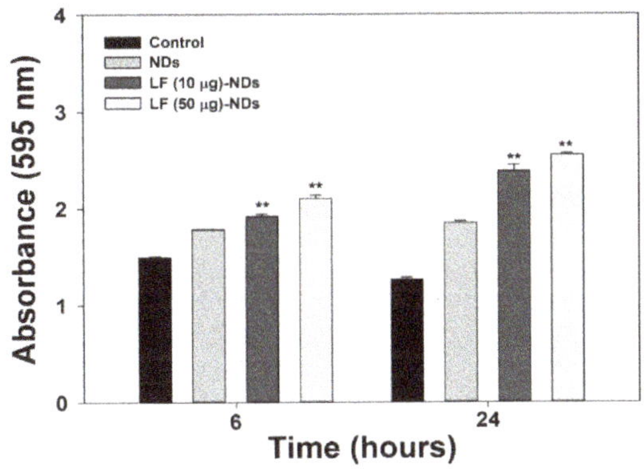

Figure 6. Cell viabilities of MC3T3-E1 cells treated with NDs with or without LF at 6 and 24 h after pre-treating under 300 μM H_2O_2 condition. The error bars represent mean ± standard deviation ($n = 4$). p value is a comparison between LF-conjugated NDs and NDs. ** $p < 0.01$.

3.6. Levels of Pro-Inflammatory Cytokines in Cell Supernatants of LPS-Induced MC3T3-E1 Cells Treated with LF-NDs

Figure 7 shows the levels of pro-inflammatory cytokines, including IL-1β and TNF-α, in cell supernatant secreted by LPS-induced MC3T3-E1 cells in the presence or absence of each sample at pre-designated time intervals of 2, 6, 24, 72 and 120 h. The levels of pro-inflammatory cytokines in MC3T3-E1 cells treated with LPS significant increased compared to untreated cells in a time-dependent manner. However, the treatment of MC3T3-E1 cells with LPS including NDs with or without LF reduced cytokines of IL-1β and TNF-α compared with those of MC3T3-E1 cells with LPS treatment. When comparing NDs with LF and NDs, we noted significant suppression of IL-1β and TNF-α. Moreover, the cytokines IL-1β and TNF-α decreased significantly in MC3T3-E1 cells treated with LF (50 μg)-NDs compared to LF (10 μg)-NDs.

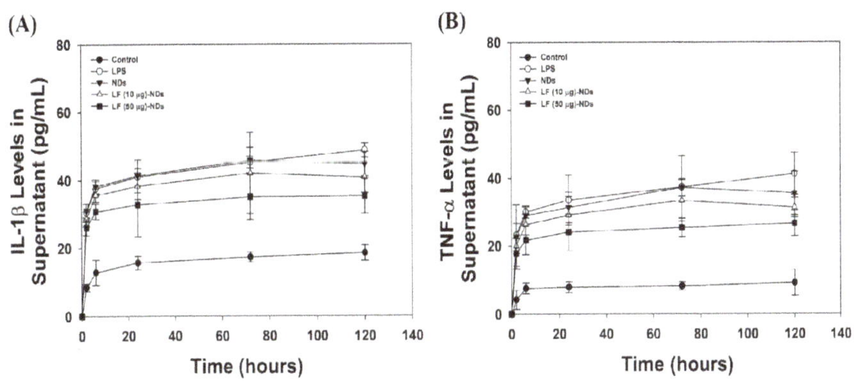

Figure 7. (**A**) IL-1β and (**B**) TNF-α levels in cell supernatant secreted by LPS-stimulated cells treated with NDs with or without LF at 2, 6, 24, 72 and 120 h by ELISA.

3.7. Alkaline Phosphatase (ALP) Activity and Calcium Deposition

To assess whether NDs with or without LF are effective for the differentiation of MC3T3-E1 cells, we measured ALP activity at 3 and 7 days. As shown in Figure 8A, the in vitro ALP activities of MC3T3-E1 cells treated with all test samples increased gradually in a time-dependent manner. The addition of NDs with LF significantly promoted ALP activity compared with NDs at 3 and 7 days. As expected, MC3T3-E1 cells treated with LF (50 μg)-NDs exhibited higher ALP activity than did cells treated with NDs and LF (10 μg)-NDs at 3 and 7 days. Generally, calcium deposition was measured as a marker of osteogenic differentiation, and upregulation of calcium deposition is a major event that occurs during late time points of osteogenesis [20,32]. Figure 7B shows the in vitro amounts of calcium deposited by MC3T3-E1 cells treated with different samples for different culture times. The amount of calcium deposited increases as incubation time intervals increase up to 21 days in all experimental groups. Calcium deposition by MC3T3-E1 cells treated with NDs containing LF was markedly higher than for MC3T3-E1 cells treated with NDs at 7 and 21 days. Moreover, significant differences in the amount of calcium deposited by MC3T3-E1 cells treated with LF (50 μg)-NDs vs. LF (10 μg)-NDs were observed at 7 and 21 days. These results supported that LF-NDs promoted the osteoblastic differentiation of MC3T3-El cells.

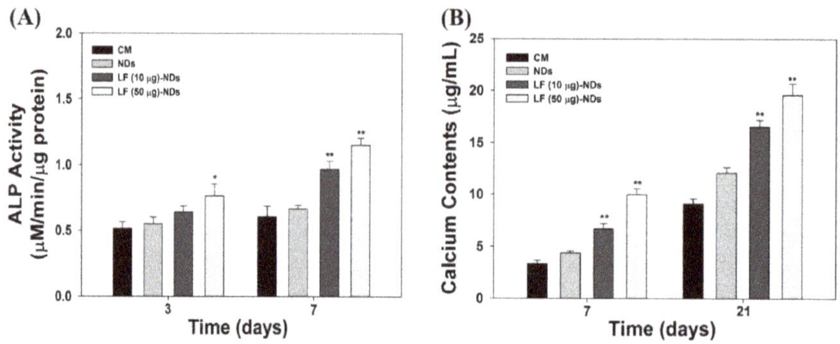

Figure 8. (**A**) Alkaline phosphatase (ALP) activity of MC3T3-E1 cells treated with LF (10 μg)-NDs and LF (50 μg)-NDs after 3 and 7 days of incubation; (**B**) Calcium deposition by ME3T3-E1 cells treated with NDs, LF (10 μg)-NDs and LF (50 μg)-NDs after 7 and 21 days of incubation. The error bars represent mean ± standard deviation (n = 4). p value is a comparison between LF-conjugated NDs and NDs. * $p < 0.05$ and ** $p < 0.01$.

4. Discussion

Bone tissue undergoes continuous bone remodeling throughout life, in which bone resorption and bone formation are regulated by the parallel activity of osteoblasts and osteoclasts [33,34]. The bone remodeling cycle, in which the structure of the bone is organized in regular units and the mass of the bone gains maximum resistance to mechanical forces acting on the bone, entails three stages: (1) the initiation of osteoclasts to form and reabsorb damaged bone; (2) the conversion of osteoclasts into osteoclast activity; and (3) formation when osteoclasts replace a portion of the reabsorbed bone [35]. Hormonal imbalances or aging can lead to osteoporosis through the disruption of bone resorption and balance, which eventually increases the risk of bony fracture.

The purpose of this study was to investigate whether osteogenic differentiation of MC3T3-E1 cells on anti-oxidant and anti-inflammatory LF-NDs could be improved by fabricating lactoferrin-conjugated NDs. LF-conjugated NDs were fabricated via electrostatic interactions between amine groups of LF and carboxyl groups of NDs. NDs, LF (10 μg)-NDs and LF (50 μg)-NDs were observed by TEM to be approximately 200 nm in size. The particle sizes of each sample measured by DLS were also confirmed to be about 200 nm. In addition, NDs after conjugating LF showed an increase N1s content and a decrease C1s content in comparison to those of bare NDs when measured by XPS. As previously reported, N1s contents increased on heparin-porous microspheres (Hep-PMs) and heparin-titanium (Hep-Ti) after immobilizing LF, compared with PMs or Ti [20,32]. These previous results were consistent with our results in the present study, and indicate that successful LF conjugation on carboxylated NDs may be achieved by electrostatic interactions.

ROS, such as superoxide anion (O_2^-), hydrogen peroxide (H_2O_2), and hydroxyl radical (HO•), are oxygen-containing molecules that play a detrimental role in age-related diseases because their levels increase with age or the onset of inflammation [36]. The imbalance between ROS production and antioxidant mechanisms leads to oxidative stress affecting the bone, which eventually accelerates the destruction of calcified tissue and bone resorption. Hydrogen peroxide (H_2O_2), which has strong oxidizing properties and is formed by many oxidizing enzymes, can cross the membrane and oxidize many compounds slowly, and thus is widely used to induce oxidative stress in vitro [5,37].

In order to confirm the radical scavenging activity of each samples, we conducted indirect and direct assessments in cells exposed to H_2O_2 such as DCFDA assays and cell viability assays, respectively. For determination of the scavenging activity in each sample using indirect methods, cells were pre-treated with 300 μM H_2O_2 to stimulate oxidative stress, followed by treating extracts from each sample. As observed in the fluorescence assay and CLSM images, NDs with or without LF significantly decrease the fluorescence signal and images in cells compared to cells without sample treatment at 6 and 24 h. Moreover, fluorescence signals and images in cells treated with the extracts from LF-NDs decreased compared to those from NDs in a dose- and time-dependent manner. To further estimate the scavenging activity of all test groups using direct methods, we measured cell viabilities treated with 300 μM H_2O_2 condition in the presence or absence of each test group. The cell viabilities were diminished in a time-dependent fashion due to oxidative damage of cellular components by H_2O_2 stimulation [38,39]. However, treatments with NDs and LF-NDs significantly increased cell proliferation. In addition, LF-conjugated ND groups showed much higher cell viabilities than the ND groups and extended cell proliferation in a dose- and time-dependent manner. These results indicate that LF molecules conjugated to NDs on ND surfaces can effectively counter intracellular ROS and interfere with cell suppression through oxidative damage, thus increasing cell viability and proliferation.

Oxidative stress can induce an inflammatory response through activation of redox-sensitive transcription NF-κB and is known to play an important role in inducing inflammatory responses [6,8,37]. In the early stages of bone repair, pro-inflammatory cytokines are released from the site of injury, and such cytokines can slow bone repair. As reported previously, pro-inflammatory cytokines inhibit osteogenic differentiation from MSCs and ADSCs [40,41]. Therefore, we investigated the in vitro anti-inflammatory activities of LF-NDs in inflamed cells. In order to mimic the inflammatory environment in vitro, cells were treated with LPS which is a major component of the outer membrane of Gram-negative bacteria, also known as lipoglycans and endotoxins. As previously studied, LPS-stimulated cells secreted increased amount of pro-inflammatory cytokines, such as TNF-α, IL-6 and IL-1β [42,43]. In order to determine the in vitro anti-inflammatory activities of LF-NDs, MC3T3-E1 cells were treated with LPS in the presence or absence of NDs with or without LF to induce an in vitro inflammatory environment, followed by the acquirement of supernatant secreted by cells at predetermined time points and measurements of the pro-inflammatory cytokines (IL-1β and TNF-α) using ELISA. Treatment with LPS upregulated IL-1β and TNF-α levels, whereas cells treated with NDs with or without LF showed decreased IL-1β and TNF-α levels in a time-dependent manner. As expected, cells treated with NDs conjugating different LF concentrations showed lower IL-1β and TNF-α levels than did cells treated with bare NDs, due to the presence of LF, which is known for inhibiting the secretion of pro-inflammatory cytokines. Prior studies reported that LF inhibited pro-inflammatory cytokines including IL-1, IL-6 and TNF-α in a monocytic cell line (THP-1) stimulated by LPS [44,45]. Rasheed et al. [46] demonstrated that LF inhibited prostaglandin E_2 (PGE_2) production and cyclocoxygenase-2 (COX-2) expression in IL-1β-induced human osteoarthritis via suppression of NF-κB activation. Consistent with these studies, we found that LF-NDs suppressed pro-inflammatory cytokines, such as IL-1β and TNF-α.

The ALP activities and calcium deposition of NDs conjugating LF-treated MC3T3-E1 cells were markedly higher than for those treated with bare NDs in a dose- and time-dependent manner, because LF molecules released from NDs affected osteogenic differentiation [20,32,47]. These findings suggest that LF molecules conjugated on NDs can induce osteogenic differentiation of MC3T3-E1 cells by sustained release of LF compared with bare NDs.

Consequently, this study demonstrated that NDs as a delivery carrier effectively ferry LF into cells. Through the effective delivery of LF with pleiotropic effects such as anti-inflammatory and anti-oxidant properties, LF-NDs could exert a synergistic effect on the osteogenic differentiation of MC3T3-El cells. Therefore, we expect that the anti-inflammatory and anti-oxidant LF-NDs will be applicable to treat bone tissue regeneration.

5. Conclusions

In this study, LF-conjugated NDs were developed to investigate their effects against oxidative stress, inflammatory response and osteogenic differentiation of cells. LF-conjugated NDs were first fabricated by electrostatic interactions between amine groups of LF and carboxyl groups of NDs. LF-NDs not only effectively scavenge ROS in cells, but also protect cells in ROS environments and can significantly suppress the levels of pro-inflammatory cytokines (IL-1β and TNF-α) secreted by LPS-stimulated cells. In addition, LF-NDs induce osteogenic differentiation of MC3T3-E1 cells by enhancing ALP activity and calcium deposition via release of LF. Thus, LF-NDs exhibit superior capacities for enhanced anti-oxidant and anti-inflammatory functions as well as induced improved osteogenic differentiation of cells. LF-NDs have great potential for application in bone regeneration and disease treatment.

Author Contributions: S.E.K. and S.-H.L. conceived and designed the study; S.E.K., S.C. and J.-Y.H. performed the in vitro experiment; K.-S.S., T.-H.K. and K.P. analyzed the in vitro data; S.E.K. and K.P. wrote the paper. All authors have read and agreed to the published version of the manuscript.

Funding: This study was supported by grants from the Bio & Medical Technology Development Program of the NRF funded by the Korean government, MSIP (NRF-2017M3A9B3063640), and a National Research Foundation of Korea (NRF) grant funded by the Korean government (MSIT) (No. NRF-2017R1C1B5015063).

Conflicts of Interest: The authors declare no conflict of interest.

References

1. Zimmermann, C.E.; Borner, B.I.; Hasse, A.; Sieg, P. Donor site morbidity after microvascular fibula transfer. *Clin. Oral Investig.* **2001**, *5*, 214–219. [CrossRef]
2. Silber, J.S.; Anderson, D.G.; Daffner, S.D.; Brislin, B.T.; Leland, J.M.; Hilibrand, A.S.; Vaccaro, A.R.; Albert, T.J. Donor site morbidity after anterior iliac crest bone harvest for single-level anterior cervical discectomy and fusion. *Spine (Phila Pa 1976)* **2003**, *28*, 134–139. [CrossRef]
3. Moreau, M.F.; Gallois, Y.; Basle, M.F.; Chappard, D. Gamma irradiation of human bone allografts alters medullary lipids and releases toxic compounds for osteoblast-like cells. *Biomaterials* **2000**, *21*, 369–376. [CrossRef]
4. Lewandrowski, K.U.; Rebmann, V.; Passler, M.; Schollmeier, G.; Ekkernkamp, A.; Grosse-Wilde, H.; Tomford, W.W. Immune response to perforated and partially demineralized bone allografts. *J. Orthop. Sci.* **2001**, *6*, 545–555. [CrossRef]
5. Almeida, M.; Han, L.; Martin-Millan, M.; O'Brien, C.A.; Manolagas, S.C. Oxidative stress antagonizes wnt signaling in osteoblast precursors by diverting beta-catenin from t cell factor-to forkhead box o-mediated transcription. *J. Biol. Chem.* **2007**, *282*, 27298–27305. [CrossRef]
6. Almeida, M.; Ambrogini, E.; Han, L.; Manolagas, S.C.; Jilka, R.L. Increased lipid oxidation causes oxidative stress, increased peroxisome proliferator-activated receptor-gamma expression, and diminished pro-osteogenic wnt signaling in the skeleton. *J. Biol. Chem.* **2009**, *284*, 27438–27448. [CrossRef]
7. Almeida, M.; Martin-Millan, M.; Ambrogini, E.; Bradsher, R., 3rd; Han, L.; Chen, X.D.; Roberson, P.K.; Weinstein, R.S.; O'Brien, C.A.; Jilka, R.L.; et al. Estrogens attenuate oxidative stress and the differentiation and apoptosis of osteoblasts by DNA-binding-independent actions of the eralpha. *J. Bone Miner. Res.* **2010**, *25*, 769–781.

8. Bai, X.C.; Lu, D.; Bai, J.; Zheng, H.; Ke, Z.Y.; Li, X.M.; Luo, S.Q. Oxidative stress inhibits osteoblastic differentiation of bone cells by erk and nf-kappab. *Biochem. Biophys. Res. Commun.* **2004**, *314*, 197–207. [CrossRef]
9. Lean, J.M.; Davies, J.T.; Fuller, K.; Jagger, C.J.; Kirstein, B.; Partington, G.A.; Urry, Z.L.; Chambers, T.J. A crucial role for thiol antioxidants in estrogen-deficiency bone loss. *J. Clin. Investig.* **2003**, *112*, 915–923. [CrossRef]
10. Manolagas, S.C. From estrogen-centric to aging and oxidative stress: A revised perspective of the pathogenesis of osteoporosis. *Endocr. Rev.* **2010**, *31*, 266–300. [CrossRef]
11. Brock, J.H. The physiology of lactoferrin. *Biochem. Cell Biol.* **2002**, *80*, 1–6. [CrossRef]
12. Hu, P.; Zhao, D.; Zhao, F.; Wang, J.; Zhu, W. The effects of the combination of oral lactoferrin and iron injection on iron homestasis, antioxidative abilities and cytokines activities of suckling piglets. *Animals (Basel)* **2019**, *9*, 438. [CrossRef]
13. Maneva, A.; Taleva, B.; Maneva, L. Lactoferrin-protector against oxidative stress and regulator of glycolysis in human erythrocytes. *Z. Naturforschung C J. Biosci.* **2003**, *58*, 256–262. [CrossRef]
14. Mulder, A.M.; Connellan, P.A.; Oliver, C.J.; Morris, C.A.; Stevenson, L.M. Bovine lactoferrin supplementation supports immune and antioxidant status in healthy human males. *Nutr. Res.* **2008**, *28*, 583–589. [CrossRef]
15. Cohen, M.S.; Mao, J.; Rasmussen, G.T.; Serody, J.S.; Britigan, B.E. Interaction of lactoferrin and lipopolysaccharide (lps): Effects on the antioxidant property of lactoferrin and the ability of lps to prime human neutrophils for enhanced superoxide formation. *J. Infect. Dis.* **1992**, *166*, 1375–1378. [CrossRef]
16. Spadaro, M.; Caorsi, C.; Ceruti, P.; Varadhachary, A.; Forni, G.; Pericle, F.; Giovarelli, M. Lactoferrin, a major defense protein of innate immunity, is a novel maturation factor for human dendritic cells. *FASEB J.* **2008**, *22*, 2747–2757. [CrossRef]
17. Legrand, D.; Elass, E.; Carpentier, M.; Mazurier, J. Lactoferrin: A modulator of immune and inflammatory responses. *Cell. Mol. Life Sci.* **2005**, *62*, 2549–2559. [CrossRef]
18. Gonzalez-Chavez, S.A.; Arevalo-Gallegos, S.; Rascon-Cruz, Q. Lactoferrin: Structure, function and applications. *Int. J. Antimicrob. Agents* **2009**, *33*, 301.e1–301.e8. [CrossRef]
19. Montesi, M.; Panseri, S.; Iafisco, M.; Adamiano, A.; Tampieri, A. Effect of hydroxyapatite nanocrystals functionalized with lactoferrin in osteogenic differentiation of mesenchymal stem cells. *J. Biomed. Mater. Res. Part A* **2015**, *103*, 224–234. [CrossRef]
20. Kim, S.E.; Yun, Y.P.; Lee, J.Y.; Park, K.; Suh, D.H. Osteoblast activity of mg-63 cells is enhanced by growth on a lactoferrin-immobilized titanium substrate. *Colloids Surf. B Biointerfaces* **2014**, *123*, 191–198. [CrossRef]
21. Zhang, W.; Guo, H.Y.; Jing, H.; Li, Y.X.; Wang, X.Y.; Zhang, H.; Jiang, L.; Ren, F.Z. Lactoferrin stimulates osteoblast differentiation through pka and p38 pathways independent of lactoferrin's receptor lrp1. *J. Bone Miner. Res.* **2014**, *29*, 1232–1243. [CrossRef]
22. Bolanos, K.; Kogan, M.J.; Araya, E. Capping gold nanoparticles with albumin to improve their biomedical properties. *Int. J. Nanomed.* **2019**, *14*, 6387–6406. [CrossRef]
23. Sharma, S.; Javed, M.N.; Pottoo, F.H.; Rabbani, S.A.; Barkat, M.A.; Sarafroz, M.; Amir, M. Bioresponse inspired nanomaterials for targeted drug and gene delivery. *Pharm. Nanotechnol.* **2019**, *7*, 220–233. [CrossRef]
24. Mochalin, V.N.; Shenderova, O.; Ho, D.; Gogotsi, Y. The properties and applications of nanodiamonds. *Nat. Nanotechnol.* **2011**, *7*, 11–23. [CrossRef]
25. Neugart, F.; Zappe, A.; Jelezko, F.; Tietz, C.; Boudou, J.P.; Krueger, A.; Wrachtrup, J. Dynamics of diamond nanoparticles in solution and cells. *Nano Lett.* **2007**, *7*, 3588–3591. [CrossRef]
26. Mochalin, V.N.; Pentecost, A.; Li, X.M.; Neitzel, I.; Nelson, M.; Wei, C.; He, T.; Guo, F.; Gogotsi, Y. Adsorption of drugs on nanodiamond: Toward development of a drug delivery platform. *Mol. Pharm.* **2013**, *10*, 3728–3735. [CrossRef]
27. Zhang, Q.; Mochalin, V.N.; Neitzel, I.; Knoke, I.Y.; Han, J.; Klug, C.A.; Zhou, J.G.; Lelkes, P.I.; Gogotsi, Y. Fluorescent plla-nanodiamond composites for bone tissue engineering. *Biomaterials* **2011**, *32*, 87–94. [CrossRef]
28. Zhang, Q.; Mochalin, V.N.; Neitzel, I.; Hazeli, K.; Niu, J.; Kontsos, A.; Zhou, J.G.; Lelkes, P.I.; Gogotsi, Y. Mechanical properties and biomineralization of multifunctional nanodiamond-plla composites for bone tissue engineering. *Biomaterials* **2012**, *33*, 5067–5075. [CrossRef]
29. Parizek, M.; Douglas, T.E.L.; Novotna, K.; Kromka, A.; Brady, M.A.; Renzing, A.; Voss, E.; Jarosova, M.; Palatinus, L.; Tesarek, P.; et al. Nanofibrous poly (lactide-co-glycolide) membranes loaded with diamond nanoparticles as promising substrates for bone tissue engineering. *Int. J. Nanomed.* **2012**, *7*, 5873.

30. Ahn, G.Y.; Ryu, T.K.; Choi, Y.R.; Park, J.R.; Lee, M.J.; Choi, S.W. Fabrication and optimization of nanodiamonds-composited poly(epsilon-caprolactone) fibrous matrices for potential regeneration of hard tissues. *Biomater. Res.* **2018**, *22*, 16. [CrossRef]
31. Kratochvilova, I.; Sebera, J.; Ashcheulov, P.; Golan, M.; Ledvina, M.; Micova, J.; Mravec, F.; Kovalenko, A.; Zverev, D.; Yavkin, B.; et al. Magnetical and optical properties of nanodiamonds can be tuned by particles surface chemistry: Theoretical and experimental study. *J. Phys. Chem. C* **2014**, *118*, 25245–25252. [CrossRef]
32. Kim, S.E.; Yun, Y.P.; Shim, K.S.; Park, K.; Choi, S.W.; Suh, D.H. Effect of lactoferrin-impregnated porous poly(lactide-co-glycolide) (plga) microspheres on osteogenic differentiation of rabbit adipose-derived stem cells (radscs). *Colloids Surf. B Biointerfaces* **2014**, *122*, 457–464. [CrossRef]
33. Vacek, T.P.; Kalani, A.; Voor, M.J.; Tyagi, S.C.; Tyagi, N. The role of homocysteine in bone remodeling. *Clin. Chem. Lab. Med.* **2013**, *51*, 579–590. [CrossRef]
34. Agidigbi, T.S.; Kim, C. Reactive oxygen species in osteoclast differentiation and possible pharmaceutical targets of ros-mediated osteoclast diseases. *Int. J. Mol. Sci.* **2019**, *20*, 3576. [CrossRef]
35. Kular, J.; Tickner, J.; Chim, S.M.; Xu, J.K. An overview of the regulation of bone remodelling at the cellular level. *Clin. Biochem.* **2012**, *45*, 863–873. [CrossRef]
36. Collins, J.A.; Diekman, B.O.; Loeser, R.F. Targeting aging for disease modification in osteoarthritis. *Curr. Opin. Rheumatol.* **2018**, *30*, 101–107. [CrossRef]
37. Almeida, M.; Han, L.; Ambrogini, E.; Bartell, S.M.; Manolagas, S.C. Oxidative stress stimulates apoptosis and activates nf-kappab in osteoblastic cells via a pkcbeta/p66shc signaling cascade: Counter regulation by estrogens or androgens. *Mol. Endocrinol.* **2010**, *24*, 2030–2037. [CrossRef]
38. Valentine, J.S.; Wertz, D.L.; Lyons, T.J.; Liou, L.L.; Goto, J.J.; Gralla, E.B. The dark side of dioxygen biochemistry. *Curr. Opin. Chem. Biol.* **1998**, *2*, 253–262. [CrossRef]
39. Rather, H.A.; Thakore, R.; Singh, R.; Jhala, D.; Singh, S.; Vasita, R. Antioxidative study of cerium oxide nanoparticle functionalised pcl-gelatin electrospun fibers for wound healing application. *Bioact. Mater.* **2018**, *3*, 201–211. [CrossRef]
40. Lacey, D.C.; Simmons, P.J.; Graves, S.E.; Hamilton, J.A. Proinflammatory cytokines inhibit osteogenic differentiation from stem cells: Implications for bone repair during inflammation. *Osteoarthr. Cartil.* **2009**, *17*, 735–742. [CrossRef]
41. Bastidas-Coral, A.P.; Bakker, A.D.; Zandieh-Doulabi, B.; Kleverlaan, C.J.; Bravenboer, N.; Forouzanfar, T.; Klein-Nulend, J. Cytokines tnf-alpha, il-6, il-17f, and il-4 differentially affect osteogenic differentiation of human adipose stem cells. *Stem Cells Int.* **2016**, *2016*, 1318256. [CrossRef]
42. Kim, S.E.; Yun, Y.P.; Shim, K.S.; Jeon, D.I.; Park, K.; Kim, H.J. In vitro and in vivo anti-inflammatory and tendon-healing effects in achilles tendinopathy of long-term curcumin delivery using porous microspheres. *J. Ind. Eng. Chem.* **2018**, *58*, 123–130. [CrossRef]
43. Park, J.W.; Yun, Y.P.; Park, K.; Lee, J.Y.; Kim, H.J.; Kim, S.E.; Song, H.R. Ibuprofen-loaded porous microspheres suppressed the progression of monosodium iodoacetate-induced osteoarthritis in a rat model. *Colloids Surf. B Biointerfaces* **2016**, *147*, 265–273. [CrossRef]
44. Mattsby-Baltzer, I.; Roseanu, A.; Motas, C.; Elverfors, J.; Engberg, I.; Hanson, L.A. Lactoferrin or a fragment thereof inhibits the endotoxin-induced interleukin-6 response in human monocytic cells. *Pediatr. Res.* **1996**, *40*, 257–262. [CrossRef] [PubMed]
45. Crouch, S.P.; Slater, K.J.; Fletcher, J. Regulation of cytokine release from mononuclear cells by the iron-binding protein lactoferrin. *Blood* **1992**, *80*, 235–240. [CrossRef]
46. Rasheed, N.; Alghasham, A.; Rasheed, Z. Lactoferrin from camelus dromedarius inhibits nuclear transcription factor-kappa b activation, cyclooxygenase-2 expression and prostaglandin e2 production in stimulated human chondrocytes. *Pharmacogn. Res.* **2016**, *8*, 135–141. [CrossRef]
47. Ying, X.; Cheng, S.; Wang, W.; Lin, Z.; Chen, Q.; Zhang, W.; Kou, D.; Shen, Y.; Cheng, X.; Peng, L.; et al. Effect of lactoferrin on osteogenic differentiation of human adipose stem cells. *Int. Orthop.* **2012**, *36*, 647–653. [CrossRef]

nanomaterials

MDPI

Article

Ex Vivo Permeation of Carprofen Vehiculated by PLGA Nanoparticles through Porcine Mucous Membranes and Ophthalmic Tissues

Lídia Gómez-Segura [1,2], Alexander Parra [3], Ana Cristina Calpena-Campmany [1,4], Álvaro Gimeno [5], Immaculada Gómez de Aranda [6] and Antonio Boix-Montañes [1,4,*]

1 Department of Pharmacy and Pharmaceutical Technology and Physical Chemistry, Faculty of Pharmacy and Food Sciences, University of Barcelona, 08028 Barcelona, Spain; lidia.gose@gmail.com (L.G.-S.); anacalpena@ub.edu (A.C.C.-C.)
2 Department of Medicine and Animal Health, Autonomous University of Barcelona, 08193 Bellatera, Spain
3 Department of Veterinary Medicine and Zootechnic, Faculty of Agricultural Sciences, University of Applied and Environmental Sciences, Bogota RX22+57, Colombia; aleparra@udca.edu.co
4 Institute of Nanoscience and Nanotechnology (IN2UB), University of Barcelona, 08028 Barcelona, Spain
5 Department of Animal Research, Animal House of Bellvitge, University of Barcelona, CCiT-UB, 08907 Hospital del Llobregat, Spain; alvarogimeno@ub.edu
6 Department of Pathology and Experimental Therapeutics, Faculty of Medicine and Health Sciences, Bellvitge Campus, University of Barcelona, 08907 Hospitalet del Llobregat, Spain; igomezdearanda@ub.edu
* Correspondence: antoniboix@ub.edu; Tel.: +34-934-024-560

Received: 7 January 2020; Accepted: 10 February 2020; Published: 18 February 2020

Abstract: (1) Background: Carprofen (CP), 2-(6-chlorocarbazole) propionic acid, is used as an anti-inflammatory, analgesic and anti-pyretic agent and it belongs to the family of non-steroidal anti-inflammatory drugs (NSAIDs). CP has some adverse reactions in systemic administration; for this reason, topical administration with CP nanoparticles (CP-NPs) can be an optimal alternative. The main objective of this work is the investigation of ex vivo permeation of CP through different types of porcine mucous membranes (buccal, sublingual and vaginal) and ophthalmic tissues (cornea, sclera and conjunctiva) to compare the influence of CP-NPs formulation over a CP solution (CP-Solution). (2) Methods: The ex vivo permeation profiles were evaluated using Franz diffusion cells. Furthermore, in vivo studies were performed to verify that the formulations did not affect the cell structure and to establish the amount retained (*Qr*) in the tissues. (3) Results: Permeation of CP-NPs is more effective in terms of drug retention in almost all tissues (with the exception of sclera and sublingual). In vivo studies show that neither of the two formulations affects tissue structure, so both formulations are safe. (4) Conclusions: It was concluded that CP-NPs may be a useful tool for the topical treatment of local inflammation in veterinary and human medicine.

Keywords: nanoparticles; carprofen; solution; drug delivery system; anti-inflammatory; veterinary diseases; NSAIDs

1. Introduction

Non-steroidal anti-inflammatory drugs (NSAIDs) are some of the most commonly used medications in human and veterinary medicine. Carprofen (CP), 2-(6-chlorocarbazole) propionic acid, is used as an anti-inflammatory, analgesic and anti-pyretic agent [1,2]. CP was licensed for systemic human use in several countries in the 1980s and for veterinary use in the 1990s. However, it was withdrawn for human use in the early 1990s for commercial reasons [3,4]. In veterinary medicine, it is still a reference anti-inflammatory and one of the most used in multiple species. The use of CP has been studied in numerous animal species (horse [5,6]; dog [7–9]; cat [10]; broiler chicken [11] and rat [12]) however there are no studies in pigs.

The main effect of CP is the inhibition of cyclooxygenase (COX), an important enzyme in the arachidonic acid cascade. This process generates other mediators involved in the inflammatory response, such as prostaglandins and thromboxanes [13]. In veterinary, it is prescribed as a solution for injection or oral tablets for the relief of pain and inflammation associated with osteoarthritis and for the control of postoperative pain associated with soft tissue and orthopedic surgeries in dogs [14]. In addition to osteoarthritis [15,16] and analgesia [17,18] it has been used against respiratory diseases and other pathologies in conjunction with antibiotics [19,20] and as anti-inflammatory after systemic administration [21]. However, some adverse reactions [3] are associated with its systemic administration, and there is no topical formulation commercially available [14]. For this reason, topical administration intended for local action can be an optimal alternative in both veterinary and human medicine [22], majorly concerning mucous tissues where drug permeation is easily suitable [23].

Numerous localized pathologies in pigs could be treated topically such as ophthalmic diseases that occur with inflammation: entropion, distichiasis, trichiasis, lagophthalmia, foreign bodies, wounds, dry keratoconjunctivitis, infections, blepharitis and conjunctivitis [24,25]. Many diseases related to buccal mucous inflammation can be also treated topically: irritations, stomatitis, infectious diseases, autoimmune diseases and fights [26]. Additionally, CP can also treat diseases that cause vaginal inflammation: microbial and parasitic infections, neoplasms, cervicitis, mucometer and subinvolution of placental sites [27,28].

The investigation of most of those pathologies in pigs is reasonably predictive of results in humans. Therefore, the study of anti-inflammatories in this species can provide much information for both veterinary and human medicine.

The suitability of a drug to penetrate through the mucous membrane is correlated with its physicochemical properties, as well as, the pharmaceutical formulation [29–31]. On the other hand, it is absolutely necessary the understanding of the processes, pathways and driving forces affecting the transmucosal permeation of the drug [32]. Following these premises, the best way to investigate the dosage of new drugs is in vivo studies. However, in early stages of drug development, in vivo studies are not easy to perform since there are specific ethical and regulatory considerations. Instead, ex vivo studies are an alternative for the investigation of formulations with topical action [33]. Although variations in the experimental setup of diffusion cells could affect the results, some factors (temperature, receptor medium, diffusion area, etc.) can be standardized to obtain consistent and predictive results. Similarly, the availability and acceptability of the mucosal formulations are directly related to the properties of the vehicle/vehicle used [34]. Therefore, characteristics of an ex vivo model to evaluate a new therapeutic agent with localized action on mucous membranes can be very much informative. For example, the comparison of different mucous membranes for the same drug and formulation enhances the robustness of a new animal model to predict topical absorption in humans [35]. Thus, the main goals of this work were the screening of different types of porcine mucous membranes in ex vivo permeation studies of CP, as well as the evaluation of the anti-inflammatory efficacy.

Another aspect to consider when treating a localized topical inflammation concerns formulation effect. It must be ensured that the drug reaches the site of action at therapeutic concentrations and, in addition, these concentrations are maintained for a prolonged time to reduce inflammation. In this sense, the use of nanoparticles (NPs) can be a good alternative to conventional drugs. They represent promising drug carriers for topical applications [36]. For topical treatments in swine, it is necessary to develop drug delivery systems that allow and facilitate the handling of the animal. The slow release nanoparticles such as those consisting of poly D, L-lactide-co-glycolide acid (PLGA) have the advantage that they are more durable over time and therefore do not require as many administrations as systemic conventional solution or tablets.

The main objectives of this work are the investigation of ex vivo permeation of CP through different types of porcine mucous membranes or commonly involved ophthalmic tissues to compare the influence of a PLGA NP formulation [32,36] over a mere drug solution.

2. Materials and Methods

2.1. Materials

CP was obtained from Capot Chemical (Hangzhou, China) and NPs of CP were prepared as described before [32,36]. Deionized water was obtained from a MilliQ1 Plus System lab. Ketamine (Imalgene®) of Merial-Boehringer Ingelheim (Sant Cugat del Vallés, Spain); Xylazine (Rompun®) of Bayer Hispania (Sant Joan Despi, Spain); Midazolam Gen® (Midazolam); Propofol Lipuro 1%® (Propofol); Forane® (Isoflurane), Pentothal® (Sodium thiopental) all of them from Centauro Veterinary (Barcelona, Spain). Endotracheal tubes with a low-pressure balloon, number 7 of Centauro Veterinary. All other chemicals were of analytical grade and used without further purification.

2.2. NPs of Carprofen

Polymeric NPs of CP were transferred by the team of Parra et al. [36]. Resultant NPs of the study of Parra et al. [36] were freeze-dried and sterilized for more studies. In our investigation, freeze-dried NPs were rehydrated by slow injection of purified water ensuring proper wetting. Then, the vial was gently shaken for 2 min until complete disintegration and dissolution of the content.

2.2.1. Materials of NPs

As explained in the study by Parra et al. [36], the materials used in that study were poly (D,L-lactic-co-glycolic) acid 75:25 (Resomer®RG753S, Mw 36,510 Da) and purchased from Evonik Industries (Essen, Germany). Poloxamer 188 (Lutrol®F68) was from BASF (Barcelona, Spain). HP-CD was obtained from Sigma-Aldrich (St. Louis, Missouri, USA) and CP from Capot Chemical (Hangzhou, P.R. China). Double distilled water was obtained from a MilliQ® Plus System, lab supplied. All other chemicals were of analytical grade and used without further purification.

2.2.2. Preparation of NPs

As detailed in the study of Parra et al. [36], NPs of matrixial structure (nanospheres) containing CP were elaborated by the solvent displacement technique under the optimized conditions. An organic solution of PLGA (1.98–7.02 mg/mL) in 10 mL of acetone containing CP (0.08–0.92 mg/mL) was poured into 20 mL of an aqueous solution of P188 (5.80–14.20 mg/mL) under moderate stirring and adjusted to pH 3.5. Acetone was evaporated and the NPs dispersion concentrated to 20 mL under reduced pressure (Buchi B-480, Flawil, Switzerland). The resulting NPs were collected by ultracentrifugation (3000 rpm, for 30 min. Sigma 301 K centrifuge, Barcelona, Spain), and washed twice with doubled distilled water [37]. Resultant NPs were freeze-dried and sterilized for in vivo study.

2.2.3. Physicochemical Characterization

The team of Parra et al. [36] described the physicochemical characterization of CP nanoparticles within the following parameters: mean particle size (Z-ave), polydispersity index (PI), zeta potential (ZP) and encapsulation efficiency (EE). The values of each parameter were: Z-ave, which ranged between 176.20 ± 0.36 and 250.17 ± 49.47 nm with narrow distribution; PI values < 0.09 ± 0.09. ZP values ranged from −19.10 ± 1.76 to −26.20 ± 0.46 mV. EE values varied from 74.70 ± 0.95 to 97.10 ± 1.41%.

2.3. Permeation Studies

2.3.1. Mucous Membranes

Three different types of porcine mucous were used as permeation membranes: buccal, sublingual and vagina. Additionally, three ophthalmic structures (sclera, cornea and conjunctiva) were also used to evaluate CP permeation and/or penetration. Ex vivo tissues were obtained under veterinary supervision from residual individuals of female pigs (cross Landrace × Large White, 25–30 kg), previously used in surgical university practices and according to the Ethics Committee of Animals Experimentation at the University of Barcelona. Pigs were anesthetized with intramuscular administration of ketamine (3 mg/kg), xylazine (2.5 mg/kg) and midazolam (0.17 mg/kg). After chirurgical experimentation, the animals were euthanized with an intravenous overdose of sodium thiopental. After debridation, samples were frozen to −20 °C and longitudinally cut in 700 μm slabs with a dermatome GA 630 (Aesculap, Tuttlingen, Germany) [38–40]. The integrity of the mucous tissues was verified prior to the experiment according to Section 2.4.

2.3.2. Ex Vivo Study: Franz Diffusion Cells

Ex vivo permeation study of NPs was performed in vertical Franz diffusion cells (FDC 400, Crown Glass, Somerville, NY, USA) following a previously validated procedure [41] with a diffusion area of 0.64 cm^2 for all membranes except for buccal mucous (2.54 cm^2). A dialysis membrane (Dialysis Tubing Visking, Medicell International Ltd., London, UK) was used according to Parra et al. [36].

Mucous membrane samples were placed between the receptor and donor compartments with the proximal side in contact with the receptor medium and the mucous side in contact with the donor chamber. Samples of 300 μL were placed in the donor compartment (CP-Solution and CP nanoparticles (CP-NPs)). As receptor medium, PBS at pH 7.4 was used and kept at 32 +/− 0.5 °C under continuous stirring in accordance with sink conditions. Samples of 300 μL were withdrawn from the receptor compartment at pre-selected times for 6 h and replaced by an equivalent volume of fresh PBS at the same temperature. This experiment was done in six replicates for each tissue and formulation. At the end of the experiment, samples were analyzed by HPLC–UV.

2.3.3. HPLC–UV

The HPLC–UV apparatus consisted of a Waters LC Module I plus (Waters Co., Milford, MA, USA), with an ultraviolet detector set up at 235 nm, and Millenium® software (Waters Co., Milford, MA, USA). A reverse-phase column C18, 3.9 × 150 nm packed with 5 μm, was used (Simmetry®, Waters Co., Dublin, Ireland). The mobile phase, previously filtered by a 0.45 μm nylon membrane filter (Technokroma, Barcelona, Spain) and degassed by sonication, consisted of methanol and potassium dihydrogen phosphate diluted in 1 L distilled water PH = 3 (75:25). The injection volume was 30 μL and the flow rate was 0.1 mL/min. This method has been validated in our laboratory and is specific and sensitive for the detection of carprofen. The retention times were 2.4–3.6 min as we can see in Figures 1–7 and Table 1.

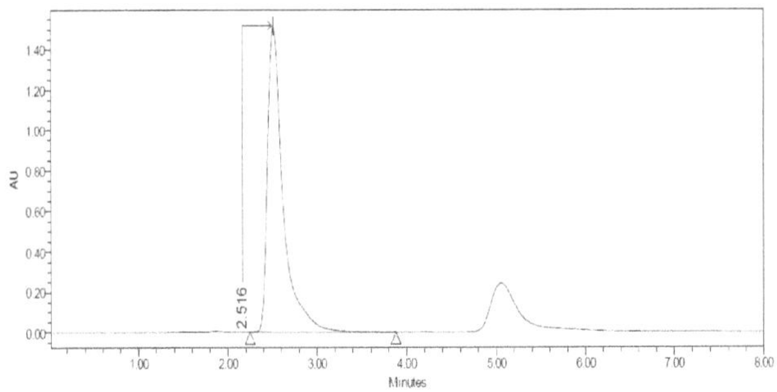

Figure 1. Chromatogram of the Carprofen (CP) validation line at concentration 100 μg/mL.

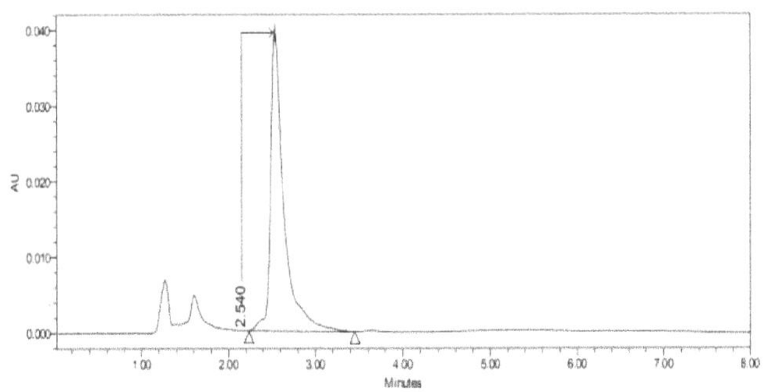

Figure 2. Chromatogram of CP nanoparticles (CP-NPs) in sclera at 6 h.

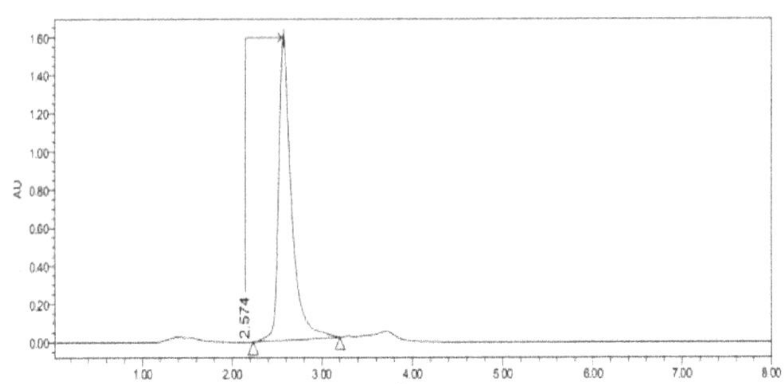

Figure 3. Chromatogram of CP-NPs in cornea at 6 h.

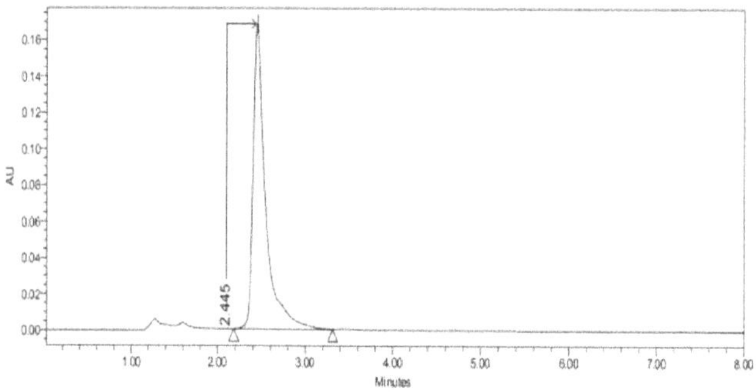

Figure 4. Chromatogram of CP-NPs in conjunctiva at 6 h.

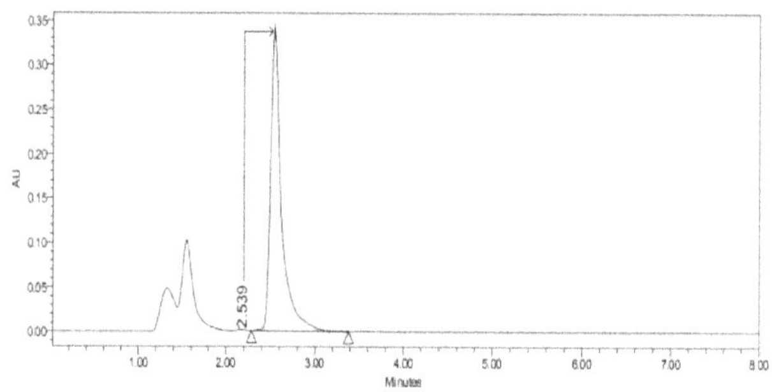

Figure 5. Chromatogram of CP-NPs in buccal mucous at 6 h.

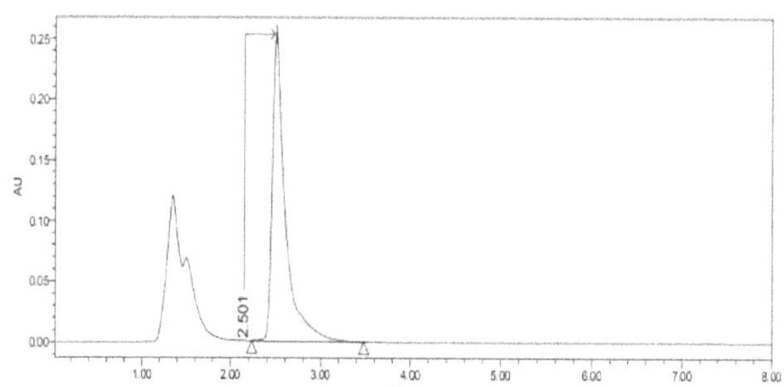

Figure 6. Chromatogram of CP-NPs in sublingual mucous at 6 h.

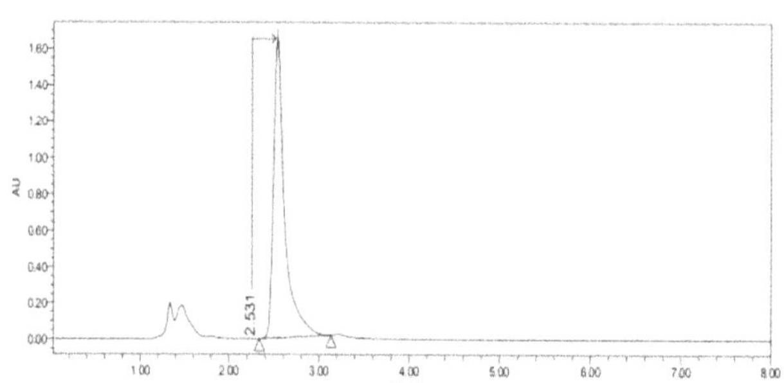

Figure 7. Chromatogram of CP-NPs in vaginal mucous at 6 h.

Table 1. Retention times (minutes) of CP-NPs in each studied tissue.

Type of Tissue	Retention Time (Minutes)
CP validation line (100 μg/mL)	2.516
Sclera	2.54
Cornea	2.574
Conjunctiva	2.445
Buccal mucous	2.539
Sublingual mucous	2.501
Vaginal mucous	2.531

2.3.4. Permeation Parameters

For the permeation assays, the flux values per unit area (*Js* in mg/h cm^2) across mucous membranes and the permeability coefficients (*Kp* in cm/h) were calculated at the steady state. Lag times (*Tl*) and *Js* values were calculated by linear regression analysis using the Graph Pad Prism® software version 5.01 (GraphPad Software Inc., San Diego, CA, USA) [42]. Stationary flux values followed the following relationship:

$$Js = Qt/A \times t \quad (1)$$

where *Qt* is the amount of CP which permeates into the receptor compartment (mg). *A* is the active cross-sectional area available for diffusion (cm^{-2}) and *t* is the time of exposure (h) per unit area

The permeability coefficient (*Kp*, cm/h) was calculated based on the relationship:

$$Kp = Js/Co \quad (2)$$

where *Js* is the flux calculated per unit area at the steady state and C_o is the drug concentration in the donor compartment.

Partition parameter (P_1) and diffusion parameter (P_2) were estimated from the following equations:

$$Tl = 1/6 \times P2 \quad (3)$$

$$Kp = P1 \times P2 \quad (4)$$

The predicted steady-state plasma concentration (*Css*) of drug that would penetrate mucous barrier after topical application, was obtained using the following equation:

$$Css = (Js \times A)/Clp \quad (5)$$

where *Css* is the plasma steady-state concentration, *Js* the flux/area obtained in this study, *A* the hypothetical area of application and *Clp* the plasmatic clearance. Calculations were addressed on the basis of a maximum application area of 1 cm^2 and *Clp* values of 2.18 L/h +/− 0.42.

The retained amount of drug in the tissue (*Qr*, μg/g/cm^2) was calculated with the following formula:

$$Qr = (Ex/Px)/A \times 100/R \quad (6)$$

where *Ex* (mcg) is the amount of drug extracted, *Px* (g) is the weight of the permeated mucous membranes, *A* (cm^2) is the active cross-sectional area available for diffusion and *R* is the percentage of recovery of the drug, obtained as described formerly [41].

2.3.5. Statistical Analysis

Nonparametric analysis was performed in these studies because drug permeation through animal tissues follows more closely a log-normal than a normal (Gaussian) distribution [43] Nonparametric *t*-test (Graph Pad Prism® 5.01, GraphPad Software Inc., San Diego, CA, USA) assessed the significance of the differences between formulations and groups of membranes. A p-value < 0.05 was accepted as significant.

2.4. In Vivo Studies

Integrity of the mucous tissues after in vivo formulations application was investigated with optical microscopy. The studies were conducted under a protocol approved by the Animal Experimentation Ethics Committee of the University of Barcelona (Spain) with ethical approval code 10619.

Thirteen female pigs (Yorkshire-Landrace) of 45–50 kg were anesthetized for 6 h with a standard sedation protocol based on IM ketamine (Imalgene®) at 3 mg/kg; IM xylazine HCl (Rompun®) at 2.5 mg/kg and IM midazolam (MidazolamGen®) at 0.17 mg/kg. Afterward, anesthesia was induced with IV propofol (Propofol Lipuro 1%®) at 2.5 mg/kg from Boehringer Ingelheim, Bayer AG, GES and BBraun, respectively. Maintenance for 6 h was achieved with inhaled isoflurane (Forane® 2%) of Centauro Veterinary administered by tracheal intubation with a low-pressure balloon, number 7.

One female pig was not administered anything (white). CP-Solution was administered to six female pigs and CP-NP was administered to six female pigs (thirteen female pigs in total). Each formulation (CP-Solution and CP-NP) was administered topically and locally in each studied tissue (conjunctiva, cornea, sclera, buccal mucous, sublingual mucous and vaginal mucous) in pigs. The same formulation (CP-Solution or CP-NP) was administered in the tissues studied in the same pig. In summary, $n = 6$ for each biological membrane and formulation and $n = 1$ for each untreated biological membrane. The reason why a single pig was used to collect untreated samples is that we did not expect to find cellular or structural changes in healthy untreated pigs. In this way, the number of animals used is reduced, as indicated in the regulation of 3R (Reduce, Replace and Refine) of animal investigation.

After the in vivo permeation phase, studied porcine mucous and ophthalmic tissues were obtained immediately after the animals were sacrificed by an intravenous overdose of sodium thiopental. Samples were collected and fixed overnight (ON) by immersion in 4% paraformaldehyde (PFA) in phosphate-buffered (PB) 20 mM, pH 7.4 and further processed for paraffin embedding. Vertical histological sections were obtained, stained with hematoxylin and eosin and mounted under a cover slip to be observed at 400× and photographed with a Leica DMD 108 optical microscope. In addition, part of the original tissue was used to calculate the in vivo retained amount (Qr) of the drug.

Determination of the Amount of Drug Remaining in the Mucous Membrane

At the end of the in vivo study, the mucous membrane specimens were used to determine the amount of retained drug. Mucous and ophthalmic samples were carefully cleaned with gauze soaked in a 0.05% solution of sodium lauryl sulfate, washed with distilled water and blotted dry with filter paper. The permeation area was then excised and weighed. Later, different tissues were perforated using a 30 G needle (BD Ultra FineTM, Beckton Dickinson, Fraga, Spain). Its CP content was extracted with methanol/buffer phosphate solution in an ultra-sonic processor for 20 min ($n = 6$ for each biological membrane and formulation). The resulting solutions were measured by HPLC–UV yielding the amount of CP retained in the membrane (Qr, mg/g.cm^2 of mucous membrane).

3. Results

3.1. Permeation Studies

The permeation studies were done six replicates for each tissue and formulation. Permeation parameters are summarized in Tables 2 and 3. Results are expressed as median, maximum and minimum values for each type of membrane and formulation. Significant differences are observed between several membranes, especially conjunctiva, buccal, sublingual and vaginal. In contrast, there are almost no significant differences between the cornea and the sclera. However, other trends have been detected and discussed. In addition, we can see in Figures 1–7 the chromatograms and CP-NPs retention times of each tissue are detailed in Table 1.

Table 2. Median, maximum and minimum values of flux (*Js*), lag time (*Tl*), P1, P2, permeability coefficient (*Kp*) and in vivo retained amount (*Qr*) of CP at 6h from the solution of CP (CP-Solution) and nanoparticles of CP (CP-NPs) through ophthalmic membranes (conjunctiva, cornea, sclera).

	CP-Solution			CP-NPs		
	Conjunctiva	Cornea	Sclera	Conjunctiva	Cornea	Sclera
Js (mcg/h)	9.4 (8.50–10.20)	5.08 (1.13–6.16)	0.9 (0.22–1.58)	4.19 ** (0.50–7.88)	1.24 (0.42–4.73)	1.01 (0.96–1.05)
Tl (h)	1.45 (1.42–1.45)	1.71 (1.51–2.33)	2.04 (1.86–2.22)	0.31 ** (0.19–1.43)	1.72 (1.31–2.44)	2.7 ** (2.66–2.73)
$P2 \times 10^1$ (h^{-1})	1.15 (0.84–1.47)	0.98 (0.72–1.10)	0.82 (0.75–0.90)	6.32 ** (3.88–8.77)	0.97 (0.68–1.27)	0.62 ** (0.61–0.63)
$P1 \times 10^2$ (cm)	17.01 (15.80–20.0)	11.91 (2.14–15.38)	2.13 (0.60–3.66)	2.18 ** (0.12–4.24)	3.34 (0.69–9.37)	3.39 (3.27–3.50)
$Kp \times 10^3$ (cm·h)	19.58 (15.40–22.0)	10.58 (2.36–12.83)	1.87 (0.45–3.29)	8.73 * (1.04–16.41)	2.58 (0.88–9.85)	2.09 (2.0–2.19)
Qr (mcg/cm²/g)	3.62 (3.61–3.63)	16.56 (16.10–17.06)	12.21 (12.17–12.26)	3.57 (2.89–4.25)	20.89 ** (18.55–23.23)	12.25 (12.16–12.34)

* p-Value < 0.05 and ** p-Value < 0.01.

Table 3. Median, maximum and minimum values of flux (*Js*), lag time (*Tl*), P1, P2, permeability coefficient (*Kp*) and in vivo retained amount (*Qr*) of CP at 6h from the solution of CP (CP-Solution) and nanoparticles of CP (CP-NPs) through mucous membranes (buccal, sublingual and vaginal).

	CP-Solution			CP-NPs		
	Buccal	Sublingual	Vaginal	Buccal	Sublingual	Vaginal
Js (mcg/h)	0.74 (0.73–0.75)	4.81 (1.37–8.24)	3.91 (3.83–3.99)	2.76 ** (1.70–3.82)	0.31 ** (0.14–0.48)	8.89 ** (5.09–12.69)
Tl (h)	1.65 (1.5–1.80)	2.77 (2.68–2.87)	3.34 (2.57–4.11)	0.8 ** (0.74–0.86)	2.09 ** (1.53–2.65)	1.75 ** (1.4–2.1)
$P2 \times 10^1$ (h^{-1})	1.02 (0.92–1.11)	0.6 (0.58–0.62)	0.53 (0.41–0.65)	2.1 ** (1.94–2.25)	0.86 ** (0.63–1.09)	0.99 ** (0.79–1.19)
$P1 \times 10^2$ (cm)	0.38 (0.35–0.41)	16.24 (4.91–27.56)	16.25 (12.82–29.67)	0.72* (0.40–1.04)	0.93 ** (0.27–1.60)	17.78 (13.35–22.21)
$Kp \times 10^3$ (cm·h)	0.39 (0.38–0.39)	10.01 (2.86–17.17)	8.15 (7.99–8.31)	1.45 ** (0.89–2.01)	0.65 ** (0.30–1.0)	18.52 ** (10.59–26.44)
Qr (mcg/cm²/g)	2.38 (2.2–2.56)	33.14 (33.09–33.18)	23.44 (20.7–26.19)	3.46 * (2.49–3.63)	29.13 ** (27.81–30.,45)	52.81 ** (49.41–56.21)

* p-Value < 0.05 and ** p-Value < 0.01.

3.2. In Vivo Studies

Histological studies of all the tissues have been carried out in order to verify if the formulations studied do affect the tissue structure and therefore, or its effect is attributable to the drug itself. In all tissues, a blank histological study (without drug), a CP-Solution histological study and a CP-NPs histological study have been performed ($n = 6$ for each biological membrane and formulation and $n = 1$ for each untreated biological membrane).

In the histological photographs of each tissue, the different layers of the untreated membrane are justified (Photos 1, 4, 7, 10, 13 and 15). The mucous membranes are composed of two parts: epithelium (A) and connective tissue (B). Inside the epithelium, the outermost part is the stratified flat keratinized epithelium (A) and the connective tissue is its own laminate (B). These two parts are separated by the basal layer. This is the basic structure of the mucous membranes and each tissue presents its own particularities detailed in the photographs.

4. Discussion

4.1. HPLC Results

Figures 1–7 show the chromatograms of the CP-NPs in each tissue studied and in Table 1 we can see different retention times of each tissue. Figure 1 represents the value of 100 μg/mL in the standard line and its retention time is 2516 min. In the following figures (Figures 2–7), the chromatograms of the samples collected at 6 h in each tissue (sclera, cornea, conjunctiva, buccal, sublingual and vaginal, respectively) are represented. As can be seen in Table 1, in all tissues the retention time it is very similar (2445–2574 min.) to the retention time of the standard line (2516 min.). Summarizing, we can say that the method used is very selective and specific for the study of CP.

4.2. Ophthalmic Tissues

As we can see in Table 2, in conjunctiva mucous the flow (*Js*) is statistically higher in the CP-Solutions (9.4 mcg/h) than in CP-NPs (4.19 mcg/h). This is interesting, since the flow is the rate of entry into the eye, preventing its residence. So, in this type of mucous, the NPs stay longer in the tissue that is the place of action of the drug. In addition, the *Tl* of the NPs is statistically lower (0.31 h) compared to the results of the CP-Solution (1.45 h). *Tl* indicates the time required to reach a steady state, therefore, the results suggested that CP-NPs are absorbed very rapidly in this tissue and had a high diffusion [44]. In the case of *Kp*, it is statistically higher in the CP-Solution (19.58 cm h) than in the CP-NPs (8.73 cm h). The *Kp* depends a lot on the formula of the drug, and in this case, the NPs have lower permeability, they stay longer in the tissue and can perform their function. Therefore, this may be an indication that NPs can have more effect for a longer time [36]. There are also significant differences between *P*1 and *P*2 between the two types of formulations. *P*1 indicates the distribution between the formulation and the tissue and *P*2 indicates the diffusion of the drug into the tissue. In this case, *P*1 is statistically higher in the CP-Solution formulation (17.01 cm) than in the CP-NPs (2.18 cm). In other words, the distribution between vehicle and membrane is greater in CP-Solutions, and therefore, it penetrates better than CP-NPs. An explanation would be the product formulation itself, since the CP-Solution is dissolved and the CP-NPs are encapsulated [45]. However, *P*2 is statistically higher in the case of CP-NPs (6.32 cm). These results show that although NPs have a shorter lag time penetrating the tissue (probably due to their formulation) [46], once inside, their diffusion and distribution capacity in the tissue is better playing a significant role in its residence inside the conjunctiva. Finally, we can see that there are no significant differences in the *Qr* of the two types of formulations (3.62 mcg/cm^2/g in CP-Solution and 3.57 mcg/cm^2/g in CP-NPs). In summary, in the conjunctive mucous, we can say that CP-NPs have advantages, since they have a lower *Js* and *Kp* and a higher *P*2 than the CP-Solution. Therefore, CP-NPs have a lower permeation capacity and a better diffusion inside the tissue.

In cornea, no statistically significant differences are observed for any parameter (Table 2), except for the *Qr*. So, both formulations act in a very similar way. Although there are no differences between *J* and *Kp*, the *Qr* of the CP-NPs (20.89 mcg/cm^2/g) is statistically greater than in the CP-Solution (16.56 mcg/cm^2/g). Therefore, we can affirm that CP-NPs are retained in the mucous and therefore have more local activity and are safer (since less drug passes into the bloodstream).

In the sclera, there are no significant differences between the *Js* and the *Kp* between formulations (Table 2). Only statistically differences are in *Tl* and *P*2. In this case, the CP-Solution has a significantly lower *Tl* (2.04 h^{-1}) than the CP-NPs (2.7 h^{-1}). In addition, the CP-Solution has a statistically higher *P*2 (0.82 cm) than the CP-NPs (0.62 cm). These two parameters suggest that in this tissue the CP-Solution has advantages over the CP-NPs, since CP-Solution takes less time to penetrate the tissue and also diffuses better inside the mucous membrane.

In brief, in the eye, we can conclude that CP-NPs have advantages in the conjunctiva and cornea. In the conjunctiva, CP-NPs take longer to penetrate, so they stay longer in the place of action. In the case of the cornea, although the two formulations act in a similar way, the CP-NPs have a better local activity and are safer. So, CP-NPs are a good option to treat locally inflammatory diseases in these two tissues.

4.3. Mucous Membranes

As we can see in Table 3, in the buccal mucous, it can be observed that J and *Kp* are statistically lower in the CP-Solution (0.74 mcg/h in CP-Solution and 0.39 cm in CP-NPs). Therefore, CP-Solution has a lower capacity for permeation and less entry into the bloodstream. However, other factors such as *Tl*, *P*1, *P*2 and *Qr* should be commented on and considered. *Tl* is statistically higher in the CP-Solution (1.65 h) than in the CP-NPs (0.8 h). In addition, *P*1, *P*2 and *Qr* are statistically significantly higher in CP-NPs. A very important factor to consider is that *Qr* in CP-NPs is statistically higher (3.46 mcg/cm^2/gr) than in CP-Solution (2.38 mcg/cm^2/gr). Therefore, we can say that CP-NPs are safer and more efficient to treat buccal mucous membranes.

In the sublingual mucous, it can be observed that the flow (*J*) is statistically lower in the CP-NPs (0.31 mcg/h) than in the CP-Solution (4.81 mcg/h) (Table 3). Therefore, as explained before, CP-NPs have a lower bloodstream entry rate, and for this reason, are more effective at the local level. In addition, *Kp* is also statistically lower in CP-NPs (0.65 cm·h) and this factor also makes it a better treatment locally in this tissue. Another factor to take into account is that the *Tl* of the CP-NPs is statistically lower (0.8 h) than that of the CP-Solution (2.77 h), and therefore, it is absorbed into the tissue more quickly. Significant differences are also observed in *P*1 and *P*2. In this case, the P1 is statistically higher in the CP-Solution (16.24 cm); but, the *P*2 is statistically higher in the CP-NPs (2.1 h^{-1}) than in the CP-Solution (0.6 h^{-1}). Therefore, CP-NPs spread faster inside the sublingual mucous. However, a negative factor is that in this tissue the *Qr* is slightly statistically higher in the CP-Solution (33.14 mcg/cm^2/g) than in the CP-NPs (29.13 mcg/cm^2/g). This factor is important, since the higher the *Qr*, the more effective and safer the drug is. However, although the *Qr* is significantly higher in the CP-Solution, the amount of *Qr* in the CP-NPs is also quite high. In summary, taking into account all the factors, we can affirm that CP-NPs are a good candidate for the local treatment of inflammatory diseases.

The last tissue to consider is vaginal mucous (Table 3). In this tissue, we can see that the flow (*J*) and *Kp* are statistically lower in the CP-Solution (3.91 mcg/h and 8.15 cm h, respectively) than in the CP-NPs (8.89 mcg/h and 18.52 cm h, respectively). Therefore, permeation into the bloodstream is lower in the CP-Solution. However, *Tl* is statistically lower in the CP-NPs (1.75 h) than in the CP-Solution (3.34 h). Then, CP-NPs reach a steady state faster than CP-Solution. In addition, *P*2 is statistically higher in CP-NPs (0.99 h^{-1}) than in CP-Solution (0.53 h^{-1}). So, the diffusion of CP-NPs within the tissue is better. Finally, we can see that the *Qr* in the CP-NPs is much higher (52.81 mcg/cm^2/g) than in the CP-Solution (23.44 mcg/cm^2/g). Therefore, the amount of drugs retained in vaginal tissue is higher in CP-NPs, and ultimately, they are safer and more effective.

In summary, we can affirm that CP-NPs have particular advantages in these three tissues. In the sublingual mucous, CP-NPs take longer to penetrate, so they stay longer in the place of action. In addition, CP-NPs are safer and more effective in the buccal and vaginal mucous. So, CP-NPs are a good option to treat locally inflammatory diseases in these three tissues.

4.4. In Vivo Studies

When reviewing histological photos, it is important to look at and compare several parameters. The most important part to analyze is the areas that have been in contact with the drug. Therefore, we will focus on the outermost area of the mucous (stratified flat keratinized epithelium and own laminate). On the other hand, it is also important to verify that the cellular structures that make up each tissue do not show alterations [47,48]. Then, we analyze the results in each tissue studied.

In buccal mucous (Figure 8) we can see that the outermost part of the tissue (A), and therefore which has been in contact with the drug, is intact in all three tissues and in addition no cell changes are observed. It can be seen that the thickness of the epithelium is different in each photo. This is due to individual variability of the animals and that there are no structural changes caused by the drugs.

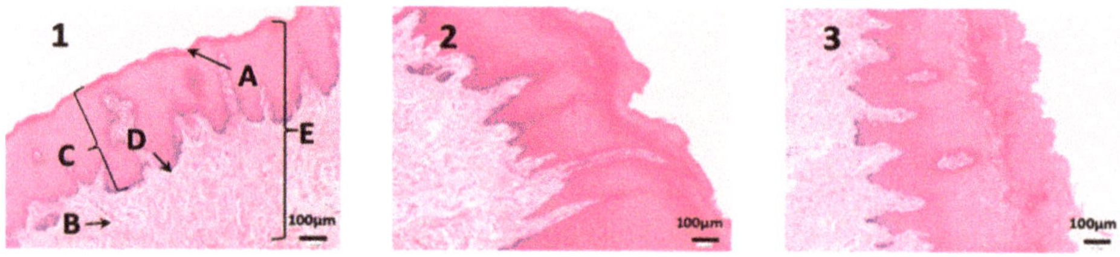

Figure 8. Photo 1: Histological image of untreated buccal mucous observed at 400×; Photo 2: Histological image of buccal mucous treated with CP-Solution observed at 400×; Photo 3: Histological image of buccal mucous treated with CP-NPs observed at 400×. (A) Stratified flat keratinized epithelium; (B) own laminate. (C) dermal papilla; (D) basal layer and (E) buccal mucous.

Sublingual mucous tissue is a much more muscular tissue than the rest (C) and with much more collagen (D) (Figure 9). However, it is important to highlight that the outermost part of the tissue (A) and cells do not show alterations, and therefore, drugs have not altered the tissue structure.

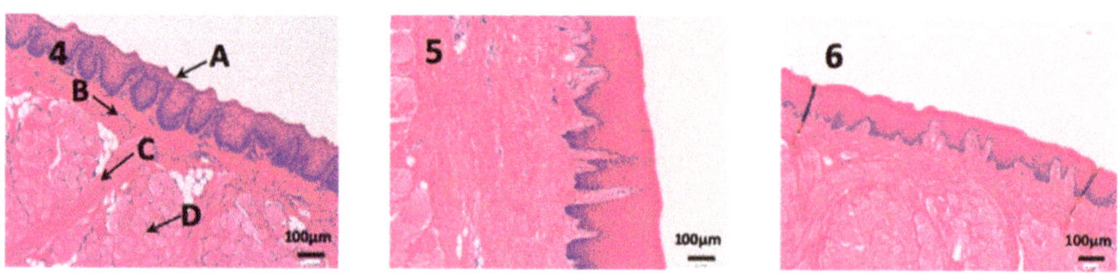

Figure 9. Photo 4: Histological image of untreated sublingual mucous observed at 400×; Photo 5: Histological image of sublingual mucous treated with CP-Solution observed at 400×; Photo 6: Histological image of sublingual mucous treated with CP-NPs observed at 400×. (A) Stratified flat keratinized epithelium; (B) own laminate. (C) muscle and (D) collagen fibers.

We can see vaginal mucous tissue presents undulations (Figure 10). This is because the pig's vaginal mucous has invaginations to facilitate intercourse with males. In addition, we can observe differences in the thickness of the epithelium (A) between each photo. These differences are due to the phase of the estrous cycle of the female pig. Depending on the estrous phase in which a female pig is, the thickness of the epithelium is different. Finally, it should be noted that the epithelium (A) is not altered in any of the three photographs.

We can see the structure of the cornea in Figure 11. It is a very fragile and fine tissue. The epithelium (A) and the own laminate (B) are separated by a thin layer called the Bowman membrane (C). As seen in photographs 11 and 12, its structure is not affected and the epithelium (A) and cells are not altered by any drug.

As we can see in Figure 12, the conjunctiva is a very vascularized tissue with oil glands (C). In addition, we can find hair follicles that belong to the eyelashes (D). As seen in photographs eight and nine, neither the epithelium (A) nor the own laminate (B) is affected by any treatment.

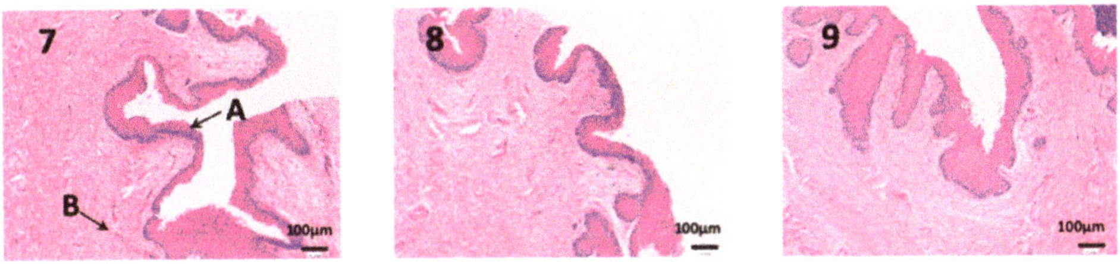

Figure 10. Photo 7: Histological image of untreated vaginal mucous observed at 400×; Photo 8: Histological image of vaginal mucous treated with CP-Solution observed at 400×; Photo 9: Histological image of vaginal mucous treated with CP-NPs observed at 400×. (A) Stratified flat keratinized epithelium; (B) own laminate.

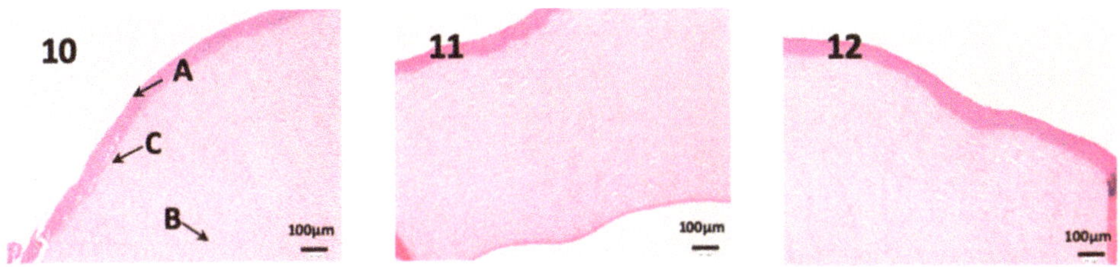

Figure 11. Photo 10: Histological image of untreated cornea observed at 400×; Photo 11: Histological image of cornea treated with CP-Solution observed at 400×; Photo 12: Histological image of cornea treated with CP-NPs observed at 400×. (A) Stratified flat keratinized epithelium; (B) own laminate. (C) Bowman's membrane.

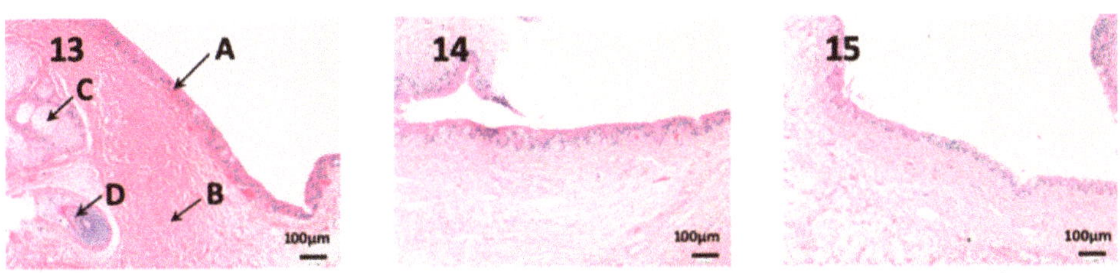

Figure 12. Photo 13: Histological image of untreated conjunctive mucous observed at 400×; Photo 14: Histological image of conjunctive mucous treated with CP-Solution observed at 400×; Photo 15: Histological image of conjunctive mucous treated with CP-NPs observed at 400×. (A) Stratified flat keratinized epithelium; (B) own laminate. (C) oil gland and (D) hair Follicle.

In this last figure, we can see the structure of the sclera (Figure 13). This tissue is similar to the cornea, it is very thin, fragile and delicate. The episclera (A) is the outermost layer and its function is to facilitate the sliding of the eyeball with the rest of the eye structures. The stroma (B) is formed by collagen fibers, and finally, the innermost layer is the fusca lamina (C), and it contains abundant blood vessels. As we can see in photographs 14 and 15, no structural part of the sclera is affected by any pharmacological treatment.

Definitely, no histopathological or significant structural changes were observed between control and treated samples. The epithelial cells and the connective tissue beneath showed normal morphology and distribution in the histological mucosal samples analyzed. These results strongly suggest that CP-Solution and CP-NPs do not affect the cellular and tissular morphology and organization in locally in vivo treatment.

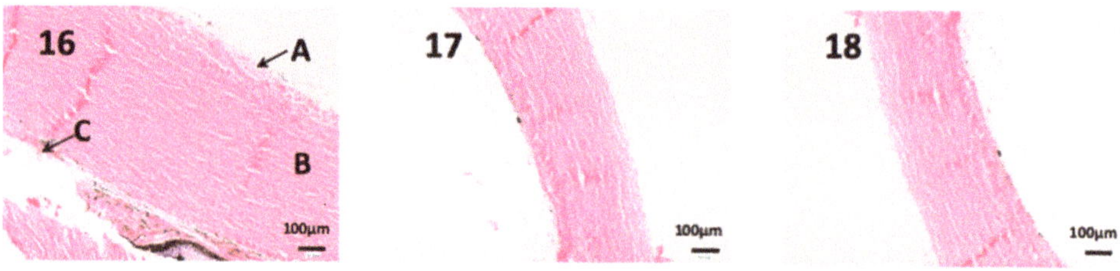

Figure 13. Photo 16: Histological image of untreated sclera observed at 400×; Photo 17: Histological image of sclera treated with CP-Solution observed at 400×; Photo 18: Histological image of sclera treated with CP-NPs observed at 400×. A: Stratified flat keratinized episclera; (B) stroma (collagen fibers). (C) fusca laminate (blood vessels).

5. Conclusions

In conclusion, the results show that CP-NPs have advantages in most tissues. They are more effective and safer than the CP-Solution and do not alter the tissue structure. This presents great possibilities for the local treatment of many inflammatory diseases in pigs or humans. In this approach, the side effects of NSAIDs will be minimized. However, additional studies are required to formulate a pharmaceutical presentation that will be easier to apply in pigs to facilitate its administration and animal management.

Author Contributions: Conceptualization, A.C.C.-C. and Á.G.; methodology, L.G.-S., A.P. and I.G.d.A.; software, L.G.-S.; validation, L.G.-S., A.P. and A.C.C.-C.; formal analysis, L.G.-S.; investigation, L.G.-S. and A.P.; resources, Á.G. and A.C.C.-C.; data curation, L.G.-S.; writing—original draft preparation, L.G.-S.; writing—review and editing, A.B.-M., A.C.C.-C. and L.G.-S.; visualization, A.B.-M., A.C.C.-C. and L.G.-S.; supervision, A.B.-M., A.C.C.-C. and Á.G. All authors contributed to the article. All authors have read and agreed to the published version of the manuscript.

Funding: This research received no external funding.

Acknowledgments: We thank all technicians and workers of the Animal House of Campus Bellvitge CCiT-UB, Joan Blasi and Alba Gómez Segura for technical assistance. The authors would like to thank the University of Barcelona for the financial support to cover the cost of open access publication.

Conflicts of Interest: The authors declare no conflicts of interest.

References

1. Yü, T.-F.; Perel, J. Pharmacokinetic and Clinical Studies of Carprofen in Gout. *J. Clin. Pharmacol.* **1980**, *20*, 347–351. [CrossRef] [PubMed]
2. Griswold, D.E.; Adams, J.L. Constitutive cyclooxygenase (COX-1) and inducible cyclooxygenase (COX-2): Rationale for selective inhibition and progress to date. *Med. Res. Rev.* **1996**, *16*, 181–206. [CrossRef]
3. Kerr, A.C.; Muller, F.; Ferguson, J.; Dawe, R.S. Occupational carprofen photoallergic contact dermatitis. *Br. J. Dermatol.* **2008**, *159*, 1303–1308. [CrossRef] [PubMed]
4. *Committe for Veterinary Medicinal Products Carpronfen Summary Report*; EMEA/MRL/042/95-FINAL; The European Agency for the Evaluation of Medicinal Products: London, UK, 1999.
5. Armstrong, S.; Tricklebank, P.; Lake, A.; Frean, S.; Lees, P. Pharmacokinetics of carprofen enantiomers in equine plasma and synovial fluid—A comparison with ketoprofen. *J. Vet. Pharmacol. Ther.* **1999**, *22*, 196–201. [CrossRef] [PubMed]
6. Lees, P.; Landoni, M.F.; Ciraudel, J.; Toutain, P.L. Pharmacodynamics and pharmacokinetics of nonsteroidal anti-inflammatory drugs in species of veterinary interest. *J. Vet. Pharmacol. Ther.* **2004**, *27*, 479–490. [CrossRef] [PubMed]
7. Lascelles, B.D.X.; Cripps, P.J.; Jones, A.; Waterman-Pearson, A.E. Efficacy and kinetics of carprofen, administered preoperatively or postoperatively, for the prevention of pain in dogs undergoing ovariohysterectomy. *Vet. Surg.* **1998**, *27*, 568–582. [CrossRef]

8. Grisneaux, E.; Pibarot, P.; Dupuis, J.; Biais, D. Comparison of ketoprofen and carprofen administered prior to orthopedic surgery for control of postoperative pain in dogs. *J. Am. Vet. Med Assoc.* **1999**, *215*, 1105–1110.
9. Slingsby, L.S.; Waterman-Pearson, A.E. Analgesic effects in dogs of carprofen and pethidine together compared with the effects of either drug alone. *Vet. Rec.* **2001**, *148*, 441–444. [CrossRef]
10. Slingsby, L.S.; Waterman-Pearson, A.E. Comparison between meloxicam and carprofen for postoperative analgesia after feline ovariohysterectomy. *J. Small Anim. Pract.* **2002**, *43*, 286–289. [CrossRef]
11. McGeown, D.; Danbury, T.C.; Waterman-Pearson, A.E.; Kestin, S.C. Effect of carprofen on lameness in broiler chickens. *Vet. Rec.* **1999**, *144*, 668–671. [CrossRef]
12. Roughan, J.V.; Flecknell, P.A. Behaviour-based assessment of the duration of laparotomy-induced abdominal pain and the analgesic effects of carprofen and buprenorphine in rats. *Behav. Pharmacol.* **2004**, *15*, 461–472. [CrossRef] [PubMed]
13. Chandrasekharan, N.V.; Simmons, D.L. The cyclooxygenases. *Genome Biol.* **2004**, *5*, 241. [CrossRef] [PubMed]
14. Food and Drug Administration. FDA Approved Animal Drug Products Database. Available online: https://animaldrugsatfda.fda.gov/adafda/views/#/home/searchResult (accessed on 13 June 2019).
15. Sanderson, R.O.; Beata, C.; Flipo, R.M.; Genevois, J.P.; Macias, C.; Tacke, S.; Innes, J.F. Systematic review of the management of canine osteoarthritis. *Vet. Rec.* **2009**, *164*, 418–424. [CrossRef] [PubMed]
16. Slingsby, L.S.; Jones, A.; Waterman-Pearson, A.E. Use of a new finger-mounted device to compare mechanical nociceptive thresholds in cats given pethidine or no medication after castration. *Res. Vet. Sci.* **2001**, *70*, 243–246. [CrossRef]
17. Bergmann, H.M.; Nolte, I.; Kramer, S. Comparison of analgesic efficacy of preoperative or postoperative carprofen with or without preincisional mepivacaine epidural anesthesia in canine pelvic or femoral fracture repair. *Vet. Surg.* **2007**, *36*, 623–632. [CrossRef]
18. Sidler, M.; Fouché, N.; Meth, I.; von Hahn, F.; von Rechenberg, B.; Kronen, P. Transcutaneous Treatment with Vetdrop® Sustains the Adjacent Cartilage in a Microfracturing Joint Defect Model in Sheep. *Open Orthop. J.* **2013**, *7*, 57–66. [CrossRef]
19. Elitok, B.; Elitok, Ö.M. Clinical efficacy of carprofen as an adjunct to the antibacterial treatment of bovine respiratory disease. *J. Vet. Pharmacol. Ther.* **2004**, *27*, 317–320. [CrossRef]
20. Lockwood, P.W.; Johnson, J.C.; Katz, T.L. Clinical efficacy of flunixin, carprofen and ketoprofen as adjuncts to the antibacterial treatment of bovine respiratory disease. *Vet. Rec.* **2003**, *152*, 392–394. [CrossRef]
21. Mathews, K.A. Non-steroidal anti-inflammatory analgesics for acute pain management in dogs and cats. *Vet. Comp. Orthop. Traumatol.* **1997**, *10*, 122–129. [CrossRef]
22. Mcpherson, M.L.; Cimino, N.M. Topical NSAID formulations. *Pain Med. (U.S.)* **2013**, *14*, S35–S39. [CrossRef]
23. Amores, S.; Domenech, J.; Colom, H.; Calpena, A.C.; Clares, B.; Gimeno, Á.; Lauroba, J. An improved cryopreservation method for porcine buccal mucous in Ex Vivo drug permeation studies using Franz diffusion cells. *Eur. J. Pharm. Sci.* **2014**, *60*, 49–54. [CrossRef] [PubMed]
24. Alqurshi, A.; Hanafy, A.F.; Abdalla, A.M.; Guda, T.K.; Gabr, K.E.; Royall, P.G. Ocular anti-inflammatory activity of prednisolone acetate loaded chitosan-deoxycholate self-assembled nanoparticles. *Int. J. Nanomed.* **2019**, *14*, 3679–3689.
25. Perpiñán, D.; Bargalló, F.; Grifols, J. Correction of periocular fat pad hypertrophy and entropion in a pot-bellied pig (Sus scrofa). *Clínica Vet. Pequeños Anim.* **2015**, *35*, 27–30.
26. Moreau, M.E.; Dubreuil, P.; Molinaro, G.; Chagnon, M.; Müller-Esterl, W.; Lepage, Y.; Adam, A. Expression of metallopeptidases and kinin receptors in swine oropharyngeal tissues: Effects of angiotensin I-converting enzyme inhibition and inflammation. *J. Pharmacol. Exp. Ther.* **2005**, *315*, 1065–1074. [CrossRef]
27. Takalkar, D.; Desai, N. Nanolipid Gel of an Antimycotic Drug for Treating Vulvovaginal Candidiasis—Development and Evaluation. *AAPS PharmSciTech* **2018**, *19*, 1297–1307. [CrossRef]
28. Newell-Fugate, A.E.; Lenz, K.; Skenandore, C.; Nowak, R.A.; White, B.A.; Braundmeier-Fleming, A. Effects of coconut oil on glycemia, inflammation, and urogenital microbial parameters in female Ossabaw mini-pigs. *PLoS ONE* **2017**, *12*, e0179542. [CrossRef]
29. Patel, A.; Bell, M.; O'Connor, C.; Inchley, A.; Wibawa, J.; Lane, M.E. Delivery of ibuprofen to the skin. *Int. J. Pharm.* **2013**, *457*, 9–13. [CrossRef]
30. Zhang, J.Y.; Fang, L.; Tan, Z.; Wu, J.; He, Z.G. Influence of ion-pairing and chemical enhancers on the transdermal delivery of meloxicam Enhanced penetration of meloxicam. *Drug Dev. Ind. Pharm.* **2009**, *35*, 663–670. [CrossRef]

31. Singh, P.; Roberts, M.S. Skin permeability and local tissue concentrations of nonsteroidal anti- inflammatory drugs after topical application. *J. Pharmacol. Exp. Ther.* **1994**, *268*, 144–151.
32. Parra, A.; Clares, B.; Rosselló, A.; Garduño-Ramírez, M.L.; Abrego, G.; García, M.L.; Calpena, A.C. Ex Vivo permeation of carprofen from nanoparticles: A comprehensive study through human, porcine and bovine skin as anti-inflammatory agent. *Int. J. Pharm.* **2016**, *501*, 10–17. [CrossRef]
33. Flaten, G.E.; Palac, Z.; Engesland, A.; Filipović-Grčić, J.; Vanić, Ž.; Škalko-Basnet, N. In vitro skin models as a tool in optimization of drug formulation. *Eur. J. Pharm. Sci.* **2015**, *75*, 10–24. [CrossRef]
34. Bolzinger, M.A.; Briançon, S.; Pelletier, J.; Chevalier, Y. Penetration of drugs through skin, a complex rate-controlling membrane. *Curr. Opin. Colloid Interface Sci.* **2012**, *17*, 156–165. [CrossRef]
35. Magnusson, B.M.; Walters, K.A.; Roberts, M.S. Veterinary drug delivery: Potential for skin penetration enhancement. *Adv. Drug Deliv. Rev.* **2001**, *50*, 205–227. [CrossRef]
36. Parra, A.; Mallandrich, M.; Clares, B.; Egea, M.A.; Espina, M.; García, M.L.; Calpena, A.C. Design and elaboration of freeze-dried PLGA nanoparticles for the transcorneal permeation of carprofen: Ocular anti-inflammatory applications. *Colloids Surf. B Biointerfaces* **2015**, *136*, 935–943. [CrossRef]
37. Zahr, A.S.; De Villiers, M.; Pishko, M.V. Encapsulation of drug nanoparticles in self-assembled macromolecular nanoshells. *Langmuir* **2005**, *21*, 403–410. [CrossRef]
38. Brugués, A.P.; Naveros, B.C.; Calpena Campmany, A.C.; Pastor, P.H.; Saladrigas, R.F.; Lizandra, C.R. Developing cutaneous applications of paromomycin entrapped in stimuli-sensitive block copolymer nanogel dispersions. *Nanomedicine* **2015**, *10*, 227–240. [CrossRef]
39. Netzlaff, F.; Schaefer, U.F.; Lehr, C.M.; Meiers, P.; Stahl, J.; Kietzmann, M.; Niedorf, F. Comparison of bovine udder skin with human and porcine skin in percutaneous permeation experiments. *Atla Altern. Lab. Anim.* **2006**, *34*, 499–513.
40. Seto, J.E.; Polat, B.E.; Lopez, R.F.V.; Blankschtein, D.; Langer, R. Effects of ultrasound and sodium lauryl sulfate on the transdermal delivery of hydrophilic permeants: Comparative in vitro studies with full-thickness and split-thickness pig and human skin. *J. Control. Release* **2010**, *145*, 26–32. [CrossRef]
41. Cañadas-Enrich, C.; Abrego, G.; Alvarado, H.L.; Calpena-Campmany, A.C.; Boix-Montañes, A. Pranoprofen quantification in Ex Vivo corneal and scleral permeation samples: Analytical validation. *J. Pharm. Biomed. Anal.* **2018**, *160*, 109–118. [CrossRef]
42. Sanz, R.; Calpena, A.C.; Mallandrich, M.; Gimeno, Á.; Halbaut, L.; Clares, B. Development of a buccal doxepin platform for pain in oral mucositis derived from head and neck cancer treatment. *Eur. J. Pharm. Biopharm.* **2017**, *117*, 203–211. [CrossRef]
43. Williams, A.C.; Cornwell, P.A.; Barry, B.W. On the non-Gaussian distribution of human skin permeabilities. *Int. J. Pharm.* **1992**, *86*, 69–77. [CrossRef]
44. Raber, A.S.; Mittal, A.; Schäfer, J.; Bakowsky, U.; Reichrath, J.; Vogt, T.; Schaefer, U.F.; Hansen, S.; Lehr, C.M. Quantification of nanoparticle uptake into hair follicles in pig ear and human forearm. *J. Control. Release* **2014**, *179*, 25–32. [CrossRef] [PubMed]
45. Carneiro, P.; Morais, S.; Carmo-Pereira, M. Nanomaterials towards Biosensing of Alzheimer's Disease Biomarkers. *Nanomaterials* **2019**, *9*, 1663. [CrossRef]
46. Bamrungsap, S.; Zhao, Z.; Chen, T.; Wang, L.; Li, C.; Fu, T.; Tan, W. Nanotechnology in therapeutics: A focus on nanoparticles as a drug delivery system. *Nanomedicine* **2012**, *7*, 1953–1971. [CrossRef]
47. Bacha, W.J., Jr.; Bacha, L.M. *Color Atlas of Veterinary Histology*, 3rd ed.; Wiley&BlackWell: New York, NY, USA, 2012; pp. 122–188.
48. Zimmerman, J.J.; Karriker, L.A.; Ramirez, A.; Schwartz, K.J.; Stevenson, G.W.; Zhang, J. *Diseases of Swine*, 11th ed.; Wiley&BlackWell: New York, NY, USA, 2019; pp. 292–392.

nanomaterials MDPI

Review

Polymeric Nanoparticles for Drug Delivery: Recent Developments and Future Prospects

Belén Begines [1,*], Tamara Ortiz [2], María Pérez-Aranda [1,3], Guillermo Martínez [1], Manuel Merinero [1], Federico Argüelles-Arias [4,5] and Ana Alcudia [1,*]

1 Department of Organic and Medicinal Chemistry, Faculty of Pharmacy, University of Seville, 41012 Seville, Spain; mariapar89@gmail.com (M.P.-A.); mtnezmun@gmail.com (G.M.); lolo191995@gmail.com (M.M.)
2 Department of Normal and Pathological Cytology and Histology, Faculty of Medicine, University of Seville, 41009 Seville, Spain; tamara.ortiz.cerda@gmail.com
3 Department of Microbiology and Parasitology, Faculty of Pharmacy, University of Seville, 41012 Seville, Spain
4 Department of Medicine, Faculty of Medicine, University of Seville, 41009 Seville, Spain; farguelles1@us.es
5 Department of Gastroenterology, University Hospital Virgen Macarena, University of Seville, 41009 Seville, Spain
* Correspondence: bbegines@us.es (B.B.); aalcudia@us.es (A.A.)

Received: 22 June 2020; Accepted: 16 July 2020; Published: 19 July 2020

Abstract: The complexity of some diseases—as well as the inherent toxicity of certain drugs—has led to an increasing interest in the development and optimization of drug-delivery systems. Polymeric nanoparticles stand out as a key tool to improve drug bioavailability or specific delivery at the site of action. The versatility of polymers makes them potentially ideal for fulfilling the requirements of each particular drug-delivery system. In this review, a summary of the state-of-the-art panorama of polymeric nanoparticles as drug-delivery systems has been conducted, focusing mainly on those applications in which the corresponding disease involves an important morbidity, a considerable reduction in the life quality of patients—or even a high mortality. A revision of the use of polymeric nanoparticles for ocular drug delivery, for cancer diagnosis and treatment, as well as nutraceutical delivery, was carried out, and a short discussion about future prospects of these systems is included.

Keywords: nanoparticles; nanocarriers; polymeric materials; drug-delivery systems; ocular delivery; cancer diagnosis; cancer drug-delivery systems; nutraceuticals

1. Introduction

The complexity of certain diseases and the toxicity associated with some treatments increasingly demand novel routes for drug delivery. A drug-delivery system (DDS) is a formulation or device that allows the introduction of active ingredients into the body in order to improve not only their efficacy, but also their safety, by controlling the drug amount, time and release in the site of action, crossing the biologic membranes to get to the therapeutic target [1]. This includes not only therapeutic drug administration methods, but also the use of vectors to facilitate their application and diffusion into the human body. In fact, different combinations of vectors and active ingredients may allow a wide range of possibilities for personalization, depending on particular diseases and patients. The routes used to administer and deliver active substances to their target tissue are a relevant factor when treating a disease [2]. These routes may have different effects depending on how they are applied. The administration is normally systemic. Occasionally, due to the severity of the disease or the toxicity inherent to the drug, it must be applied directly to the affected organ. Figure 1 shows the different anatomic routes of administration for drug delivery currently available. All delivery routes present inconveniences when delivering a formulation. As previously mentioned, the potential toxicity inherent to active ingredients or to the high dosage needed to achieve pharmacological effect, is a common disadvantage displayed by the systemic administration routes. The oral pathway of administration, for instance, limits the use of pH-resistant or highly hydrophilic drugs to ensure the required absorption by the intestinal epithelium cells. Likewise, the invasive nature of injections was associated with a high risk of infection.

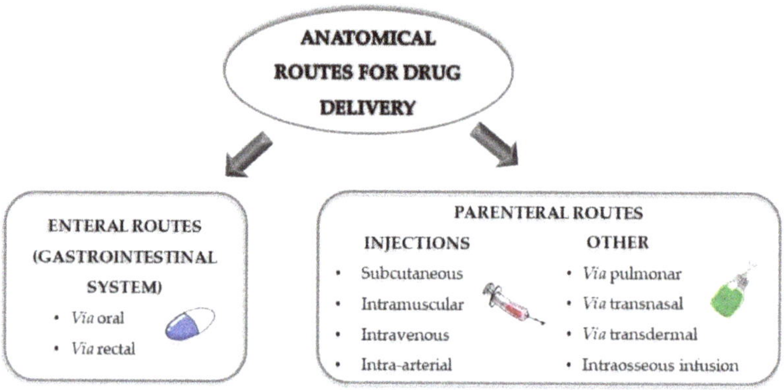

Figure 1. Classification of the different anatomic routes for drug delivery.

As stated above, to minimize the risks and disadvantages associated with traditional administration routes, DDSs are becoming increasingly more sophisticated [3], focusing on a better controlled release, maintaining therapeutic efficacy and the active ingredient targeting to the specific site of action, thus avoiding systemic release of the active substance. In this sense, nanotechnology is gaining high relevance, as it could potentially solve some of the issues associated with the above-mentioned traditional administration routes. The bioavailability refers to the portion of the bioactive compound absorbed in the body entering systemic circulation and performing functions. In general, nanoparticles (NPs) could be optimized to improve the drug bioavailability, either by increasing their absorption through enhanced solubility or by facilitating their passage through the biologic membranes [4]. Drug release could also be controlled and maintained at therapeutic levels, by adjusting the composition of the nanoparticulate system. They could even facilitate the combined therapy by the incorporation of more than one active ingredient. The progress in biologic therapies or immunotherapies has been promoted by the advances in nanotechnology, due to the fact that it allows a better administration of gen- or protein-based drugs. Functionalization of the NPs allows the recognition of the specific site of action, avoiding high systemic concentrations and reducing side effects. This property has been very useful in the diagnosis field by combining the specific targeting with the transport and release of a contrast agent [5].

According to the previously described characteristics of a nanoparticulate delivery structure, investigations about the use of different materials as nanocarrier precursors are an essential requirement for the improvement of the applicability and results achieved by these systems. These precursors should meet some requisites such as biocompatibility, biodegradability and non-immunogenicity [6]. Polymers are macromolecules formed by the covalent union of one or different sort of units, named monomers, to constitute a linear or branched chain. These monomers may possess any structure, as long as they have at least two functional groups where they can react with another monomer. Ideally, selecting the right kind of monomer/s, a polymer could be prepared to attain specific properties. Polymers are not only a special type of material that may encompass all the above-mentioned characteristics, but also, the great synthetic versatility they exhibit allows the researcher to customize them according to the requirements or final aims. In order to accomplish certain properties, polymeric tailoring could be carried out directly on biopolymers by chemical derivatization [7,8]. Another option is the preparation of synthetic polymers from their corresponding monomers which can lead to a large range of structures and applications [9–12]. These are the reasons why polymeric materials are gaining great relevance in nanotechnology in general and are being used as NP precursors for DDSs.

When considering the preparation of polymeric NPs, the use of surfactants may be a requirement. Surfactants are amphiphilic organic molecules that can self-assemble in solution. Most used surfactants are composed by a hydrocarbon chain (hydrophobic section) bound to an ionic functional group (forming the so-called cationic surfactants, such as benzalkonium chloride or tetramethylammonium hydroxide or anionic surfactants, like docusate or sodium laurate). Non-ionic surfactants can also be found, in which the amphiphilic character is generated by the union of hydrophobic and hydrophilic molecules (e.g., ethoxylated amines, alkyl and nonyl-phenol ethoxylates) [13]. Low molecular weight polymers could act as surfactants too, specially block copolymers (e.g., Pluronic F127 or Pluronic P123) [14]. In general, they are commonly included in the nanocarrier formulation as stabilizer agents and may be crucial to obtain a well-structured nanosystem, stabilizing the dispersion during nanoemulsion procedures. Some of the advantages of the stabilizers are to decrease the surface tension of NPs and increase affinity with lipidic structures [15]. Some surfactants have also demonstrated a significant reduction of the mean NPs diameter and also a double action as a cryoprotectant agent [16]. Studies of pharmacokinetics and biodistribution showed increased retention of the drug in the body and accumulation in the target tissue, prolonged time in the blood circulation along with a decreasing nephrotoxicity, hepatotoxicity, lower cardiovascular effects and reduced uptake of macrophage when surfactant surface-modified NP systems are used [17,18]. Multidrug resistance (MDR) mediated by the human ATP-binding Cassette (ABC) transporter superfamily such as *P*-glycoprotein (*P*-gp/ABCB1), multidrug resistance-associated protein 2 (MRP2/ABCC2) and breast cancer resistance protein (BCRP/ABCG2) have been recognized as the main obstacle against efficacy towards multiple chemotherapeutic agents [19]. Both organic and inorganic NPs have been demonstrated to inhibit the MDR. The effects of organic NPs are caused by several excipients, such as surfactants and polymers [20].

This review focuses on the recent advances of the most promising applications of nanoparticulate polymeric formulations as nanocarriers, considering principally those used to treat diseases with a considerable morbidity, a notable reduction in the patient's quality of life or even an important mortality.

2. Polymeric Nanocarriers for Ocular Drug Delivery

In 2019, the World Health Organization (WHO) estimated that at least 2.2 billion people had vision impairment or blindness, of whom, at least one billion cases could have been prevented with the appropriate treatment [21]. Figure 2 shows the incidence rate of the most conventional eye diseases and their corresponding proportions, throughout the world. It is evident that almost half of global blindness or vision impairment could be avoided with the correct treatment. Developing novel and efficient mechanisms for ocular therapy is a current need.

Drug delivery to the eye entails great challenges due to the actual anatomy and physiology of the organ. Two areas can be differentiated: anterior and posterior segments. The anterior segment—composed of aqueous humor, conjunctiva, cornea, iris, ciliary body and lens—is located at the front of the eye, which is readily accessible, making topical instillation of eyedrops the standard method for drug administration [22]. The posterior segment—composed of the choroid, neural retina, optic nerve, retinal pigment epithelium, sclera and vitreous humor—suffers from a low bioavailability at the site of action, due to the reduced period of time that the drug remains inside the ocular globe. Lacrimation, reflex blinking, tear-film turnover or nasolacrimal duct drainage cause a fast elimination of the drug-containing eyedrops. Formulations also need to penetrate different barriers before reaching the posterior segment. This drastically reduces the number of drugs that can achieve therapeutic effects to corticosteroids and cyclooxygenase inhibitors of higher potency [23]. Another option for drug delivery is the administration via systemic route. The presence of the blood retinal barrier drastically reduces the drug access, thus requiring a high dose to reach therapeutic effects. This, in turn, may cause systemic side effects [24]. An alternative for achieving adequate levels of active substance in the action site is directly injecting the drug. This process is associated with high costs, low patient adherence to treatment and elevated risk of injection-related complications. Implementation of intravitreal implants can also be an option for a sustained drug delivery in the posterior segment of the ocular globe, but as in the previous case, it requires numerous injections or even surgery, with the consequent risks [25]. To overcome these limitations, nanotechnological systems for ocular therapy are currently gaining high significance, especially those derived from biodegradable polymers. These systems are mainly designed to achieve an effective dose in the action site, either by improved formulation solubility properties, enhanced bioavailability, targeted delivery, sustained release or a lengthier shelf life [26]. Although different nanosystems for drug delivery to the eye can be found in literature, for example, liposomes [27], nanoemulsion [28], nanodiamonds [29], nanocrystals [30] or inorganic NPs [31], this review focuses on polymer-derived NPs. Some examples of the most recent research regarding polymeric nanoparticulate materials for ocular drug delivery, including micelles, dendrimers, cyclodextrins and polymeric vesicles, will be explored, all of them administered via ophthalmic drops and intraocular injections.

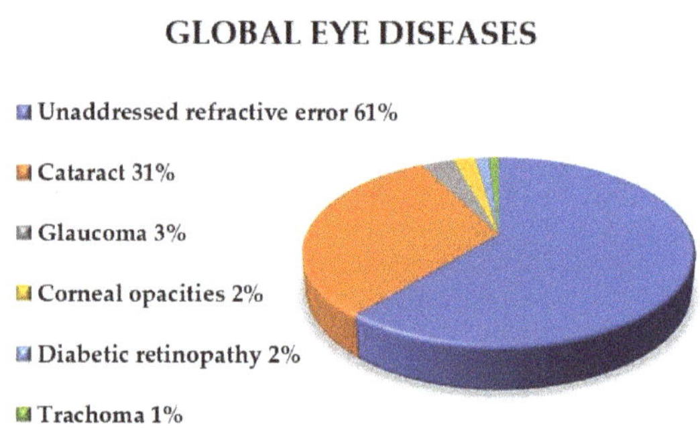

Figure 2. Incidence rate of the most conventional eye diseases, according to the WHO (2019) [21].

2.1. Micelle Nanocarriers for Ocular Delivery

Most commonly used polymers for the synthesis of micelles as DDSs is the poly(lactic-co-glycolic acid) (PLGA), due to its high biocompatibility and biodegradability. It was approved for clinical use in 1989, by the US Food and Drug Administration (FDA) [32]. It has been tested for the sustained delivery of different drugs formulated with polyvinyl alcohol (PVA). Varshochian et al. [33] used the PLGA/PVA system to prepare bevacizumab-loaded micelles for the treatment of ocular neovascularization. This molecule has widely demonstrated its effect in the treatment of retinal and choroidal neovascularization, however, its short half-life in vitreous humor requires frequent intravitreal injections. Surprisingly, bevacizumab-containing micelles provided a sustained release, and the drug concentration in vitreous humor endured above 500 ng/mL, the minimum concentration that completely blocks the vascular endothelial growth factor-induced angiogenesis, for about two months. Dexamethasone, a powerful anti-inflammatory, was also delivered in this type of NP. Ryu et al. [34] prepared dry tablets, administered with a specific preocular applicator, containing dexamethasone-loaded micelles incorporated in a alginate matrix. They demonstrated that the use of this system could increase 2.6-fold the ocular drug bioavailability when compared to Maxidex®, a dexamethasone-based eyedrop formulation, which is commercially available. PLGA/PVA NPs were also used to deliver fenofibrate, showing therapeutic effects on diabetic retinopathy and neovascular age-related macular degeneration. Qiu et al. [35] proved that the use of these micelles improved retinal dysfunctions, inhibited retinal leukostasis, diminished retinal vascular leakage and regulated the over expression of vascular endothelial growth factor (VEGF) at eight weeks after the application.

Lui et al. [36] described a similar PLGA/PVA based on micelles to treat choroidal neovascularization. In this case, the polymer system also included polyethyleneimine (PEI) to incorporate positive charges in the NP, while the medication was a combination of dexamethasone and bevacizumab, which can interact with the positive charges of PEI. These micelles demonstrated a good anti-angiogenic effect and a strong inhibitory effect on VEGF secretion from human umbilical vein endothelial cells (HUVEC). PLGA NPs have also been formulated exchanging PVA polymer for a different hydrophilic agent, such as Tween 80, poloxamer 188, Pluronic F127 or polyvinylpyrrolidone, among others. Salama et al. [37] developed a brinzolamide-loaded micelles for ocular pressure treatment, which is the most important risk factor for the appearance of glaucoma. They prepared a collection of different nanosystems utilizing a combination of two PLGA, with varied molecular weights and Tween 80, poloxamer 188 or Brij® O10. As well as showing compatibility with the ocular tissue, these micelles proved to reduce ocular pressure for up to 10 days, showing that smaller NPs were able to decrease ocular pressure longer than those with higher particle size. Another type of PLGA-based nanosystem was obtained by Pan et al. [38], who used Pluronic F127 as a hydrophilic agent. In this case, dexamethasone, a drug also used for the treatment of immunologic graft rejection, was encapsulated. Results demonstrated that these micelles prevented corneal graft rejection for at least nine weeks when administered weekly in rats. The control group did suffer from rejection, severe corneal edema, opacity and neovascularization in less than four weeks when dexamethasone in solution was applied. In a different approach, Li et al. [39] utilized polyvinylpyrrolidone to develop PLGA NPs containing bevacizumab as medication against age-related macular degeneration. This system showed a sustained drug-release for over 91 days, although this liberation could be adjusted by modifying the drug/polymer ratio.

Another widely used polymer for the preparation of micelles for ocular delivery is chitosan (CH), a natural hydrophilic cationic polysaccharide. This polymer presents excellent mucoadhesion and penetration properties that make it ideal for drug delivery in mucosa and ophthalmic areas. In this sense, Ameeduzzafar et al. [40] prepared CH-based micelles using sodium tripolyphosphate (TPP) as a crosslinking agent, for the delivery of levofloxacin for ocular infection treatment. This system proved to be biocompatible for topical ophthalmic use, to have a retention time in the ocular area longer when compared with a levofloxacin solution and to reduce corneal clearance and naso-lachrymal drainage. Bevacizumab was also encapsulated in this particulate formulation by Badiee et al. [41]. Drug-loaded CH NPs were later embedded in a hyaluronic acid ocular implant. Although in vivo experiments are not reported, in vitro studies displayed a sustained drug release over two months. Silva et al. [42] developed a similar CH/TPP system—but including hyaluronic acid—another mucoadhesive natural polymer that can react with cell-surface receptors, such as CD44. They encapsulated ceftazidime, a very unstable antibiotic, for the treatment of eye infections such as keratitis. Their studies demonstrated that this nanoformulation presented the required physicochemical and pharmaceutical characteristics for topical eye administration. It was also able to preserve the antibacterial activity while having relevant mucoadhesive properties, by interacting with mucin, an essential condition to improve the residence time in the ocular globe. Some authors have used a different crosslinking agent, instead of TPP, such as sodium deoxycholate. For example, Hanafy et al. [43] used a CH/PVA system crosslinked with sodium deoxycholate to embed prednisolone for the treatment of ocular inflammation diseases. Results obtained on female guinea pigs showed that optimized NPs formulation achieved a twofold increase in the prednisolone release after 24 h when compared with the commercial micronized drug loaded gel. Different routes to obtain CH-based micelles have been explored. One of them is the functionalization with lipophilic derivatives of the CH chain through its primary amino groups. Following this concept, Xu et al. [44] designed a novel branched CH, in which stearic acid and valylvaline were introduced on the polymer main chain in different proportions. These polymeric materials were able to self-assemble encapsulating dexamethasone. Thus, NPs demonstrated access to the posterior segment through conjunctival route, showing sustained release and enhanced penetration properties. In vivo tests carried out in male rats and male New Zealand albino rabbits, exhibited similar dexamethasone levels, compared to dexamethasone-loaded hydrogenated castor oil-40/octoxynol-40 NP, a similar system to Cequa, approved by the FDA. Another option is developing hydrophilic–hydrophobic block copolymers including a CH section. In this line, Shi et al. [45] synthetized a cationic CH grafted methoxy poly(ethylene glycol)-poly(ε-caprolactone) (PEG–PCL) for enclosing diclofenac. The polymer amphiphilic character enables a self-assembly to form micelles at the same time that its positive charges can interact with the negatively charged mucin and increase the NP retention time at the site of action. This formulation showed to be nontoxic and to present enhanced penetration and retention of diclofenac compared with the drug commercial eye drops (1.4-fold higher). The diclofenac concentration in the aqueous humor of rabbits was 2.3-fold higher than that achieved when applying the commercial drug eye formulation.

PLGA and CH are probably the most investigated polymers for the preparation of nanoparticulate formulations for ocular drug delivery, albeit different systems have also been tested. For instance, cyclosporine A, a immunosuppressive agent to treat dry eye syndrome, was encapsulated by Yu et al. [46] in a set of micelles based on the methoxy PEG–PCL block copolymer, modifying the proportions of each block. Results displayed that this system could achieve a 4.5-fold increase in retention effect when compared with 0.05% cyclosporine A emulsion. Another example is the NP obtained by Tang et al. [47] who used a mixture of Tween-80 and polyoxyethylene stearate to encapsulate everolimus (40-*O*-(2-hydroxyethyl)-rapamycin), a drug administered for autoimmune diseases, such as autoimmune uveoretinitis, non-infectious uveitis, corneal neovascularization and immune-mediated rejection after corneal transplantation. They also prepared a nanosuspension containing micelles from PVA, poloxamer P407 (an amphiphilic triblock copolymer containing a hydrophobic section of poly(propylene oxide) and two hydrophilic blocks of poly(ethylene oxide)) and hydroxypropyl methylcellulose. In vivo experiments showed that both systems are promising as ophthalmologic drug carriers, although nanosuspension presented higher release, permeability and bioavailability in New Zealand white rabbits' eyes. Another possibility is the derivatization of the drug to improve its encapsulation inside the micelles. Huang et al. [48] modified triamcinolone acetonide, a drug used for eye inflammation, as its succinated derivative. Together with PEG–PCL–PEG block copolymer, they obtained a nanoformulation that demonstrated higher transcorneal drug permeability in isolated rabbit cornea, when compared to triamcinolone acetonide suspension. Enhanced therapeutic efficacy against endotoxin-induced uveitis in rabbit model was also displayed.

Some authors improved formulation efficacy by combining the use of drug-loaded micelles with specific media to enhance certain properties, mainly oriented to prolong the residence time of the formulation in the ocular globe. Wen et al. [49] prepared a NP-loaded in situ gel as dexamethasone delivery system for ocular inflammation. The drug was contained in lecithin-based micelles while the in situ gel was obtained by a mixture of poloxamers (P188 and P407). This mixture of polymers is liquid at room temperature, for an easier administration as ocular eye drops, but it turns into a gel at temperatures higher than 35 °C. This property increased the formulation residence time in the eye with the consequent increment in the drug delivery duration. Similarly, Hirani et al. [50] developed a thermoreversible gel loaded with PLGA NPs as drug-delivery system of triamcinolone acetonide for the treatment of age-related macular degeneration. Micelles were prepared using a diblock copolymer of PLGA and PEG, while the thermoreversible gel was obtained by optimization of the amounts of NPs suspensions and a PEG–PLGA–PEG triblock copolymer. Using a different approach, Yandrapu et al. [51] optimized a system in which poly(lactic acid) (PLA) NPs, coated with bevacizumab, were embedded in PLGA microparticles using supercritical CO_2. This technique has the advantage of avoiding the use of organic solvents or sonication for the preparation of the nanosystem, that could denature the quaternary structure of protein drugs, such as the above-mentioned bevacizumab. With this methodology, Yandrapu et al. [51] demonstrated in a rat model, that the delivery of this drug increased from two weeks, when administered in solution, to two months, when applied in the NP/microparticle system.

2.2. Dendrimeric Nanocarriers for Ocular Delivery

Although most of the explored polymeric nanocarriers have focused on micelle-based NP systems, dendrimeric structures have also been investigated. The archetypical dendritic polymer for the preparation of nanosystems for drug delivery to the posterior segment of the eye is based on polyamidoamines. Yang et al. [52] examined the potentiality of dendrimers prepared from a PEGylated polyamidoamine and modified with cyclic arginine–glycine–aspartate hexapeptide and penetratin, as drug carriers. They demonstrated that these functionalized NPs were present in the ocular posterior segment after more than 12 h of a non-invasive administration. A dendrimer based on a polyamidoamine was also studied by Lancina et al. [53] In this case, the dendrimeric core was derivatized with a timolol analog, a common drug used for the treatment of ocular hypertension. Their results displayed an intraocular pressure reduction of 30% in normotensive adult Brown Norway male rats, after 30 min of topical application, in addition to the absence of irritation or toxicity after one week of daily administration. Tai et al. [54] investigated the formation of the complex between polyamidoamine dendrimer and hyaluronic acid. This complex was functionalized with penetratin and loaded with antisense oligonucleotides for the treatment of ocular diseases by regulating the expression of target proteins and genes in cells. This system showed to possess an enhanced eye permeability and distribution to the ocular posterior segment, representing a promising formulation for ocular topical administration. Despite the fact that most recent scientific research dealing with dendrimeric NPs has been based on the use of a certain type of polyamidoamine polymer, some other dendrimer structures can be found in literature [55], such as polylysine (PLL) [56] or phosphorous dendrimers [57].

2.3. Other Types of Polymeric Nanocarriers for Ocular Delivery

Although they are less utilized for ocular drug administration, cyclodextrins (CDs) and polymeric vesicles (PVs) are another type of polymeric nanocarriers. CDs are a special type of cyclic oligo- or polysaccharide constituted of six or more units of glucose bound by α-1,4 glycosidic bonds [58]. This characteristic configuration entails truncated cone geometry with an outer surface presenting a hydrophilic character and an internal cavity with hydrophobic feature. This makes them a good option as DDS for hydrophobic active ingredients [59]. In addition, the polarity of these systems can be modulated depending on the number of glucose units that form the cyclodextrin and the variety of their substituents. Rodriguez-Aller et al. [60] developed a library of different cyclodextrins as nanocarriers of latanoprost, an insoluble prostaglandin F2a analog used for glaucoma treatment. Results demonstrated that the ideal candidate as DDS of latanoprost was the propylamino-β-CD, which entails an enhanced ocular tolerance and hence, better patient compliance. It showed better drug stability and availability, as well as lower eye irritation, when compared with commercial latanoprost formulation. Jansook et al. [61,62] used γ-cyclodextrin and randomly methylated β-cyclodextrin to enhance solubility of celecoxib, a non-steroidal anti-inflammatory administered for age-related macular degeneration and diabetic retinopathy. By combination of these nanoaggregates with mucoadhesive polymers, such as hydroxypropyl methylcellulose or hyaluronic acid, they obtained eye-drop formulations that demonstrated improvements in drug permeation through transcorneal and transscleral routes, with no cytotoxicity shown. In a similar approach, Lorenzo-Veiga et al. [63] prepared ocular natamycin nanocarriers. Currently, this is the only drug approved for fungal keratitis treatment. They used a combination of Soluplus and Pluronic P103 and α-cyclodextrin to generate a library of micelles and poly(pseudo)rotaxanes containing the drug. The latter were found to be the most promising candidates since they displayed good diffusion, cornea and sclera accumulation and sclera permeability coefficients.

Although vesicles possess mainly a lipidic nature, some polymer-derivatized phospholipids have been used for their preparation. In most cases, the modification was carried out by the introduction of a PEGylated unit in the vesicle outer surface, since they have demonstrated improved penetration levels, better bioavailability and reduced toxicity compared to non-modified vesicles [64]. In this sense,

Zorzi et al. [65] developed a PEGylated PV to encapsulate siRNA sequences for the treatment of ocular keratitis caused by *Acanthamoeba*. A combined therapy of siRNA-loaded PV with chlorhexidine led to a 60% reduction in corneal damage caused by this disease in a murine model. In a similar example, PEGylated vesicles were burdened with natamycin by Patil et al. [66], demonstrating a considerably higher in vitro transcorneal permeability when compared to commercial Natacyn®.

Table 1 displays a summary of the composition of the previously described polymeric NPs, the drug they encapsulated and the disease for which they are used.

Table 1. Polymeric micelles used for drug delivery, drug encapsulated and disease for which they are administered.

Type of Nanoparticle	Nanoparticle Composition	Drug Delivery	Treatment	Reference
Polymeric Micelles	PLGA/PVA	bevacizumab	Choroidal and retinal neovascularization	[33]
		dexamethasone	Ocular inflammation	[34]
		fenofibrate	Retinal dysfunctions, retinal leukostasis, retinal vascular leakage, over expression of VEGF, choroidal neovascularization	[35]
	PLGA/PVA/PEI	bevacizumab and dexamethasone	Choroidal neovascularization	[36]
	PLGA/Tween 80, poloxamer 188 or Brij®	brinzolamide	Ocular pressure	[37]
	PLGA/Pluronic F127	dexamethasone	Immunologic graft rejection	[38]
	PLGA/PVP	bevacizumab	Age-related macular degeneration	[39]
	CH/Sodium tripolyphosphate	levofloxacin	Ocular infections	[40]
		bevacizumab	Choroidal neovascularization	[41]
	CH/Sodium tripolyphosphate/hyaluronic acid	ceftazidime	Ocular infections	[42]
	CH/PVA/sodium deoxycholate	prednisolone	Ocular inflammation	[43]
	Stearic acid and valylvaline functionalized CH	dexamethasone	Ocular inflammation, retinal dysfunctions, retinal leukostasis, retinal vascular leakage, over expression of VEGF, choroidal neovascularization	[44]
	Cationic CH grafted methoxy poly(ethylene glycol)-poly(ε-caprolactone)	diclofenac	Ocular inflammation	[45]
	Methoxy poly(ethylene glycol)-poly(lactide) block copolymer	cyclosporine A	Dry eye syndrome	[46]
	Tween80/polyoxyethylene stearate	everolimus	Autoimmune uveoretinitis, non-infectious uveitis, corneal neovascularization and immune-mediated rejection	[47]
	PVA/Poloxamer P407/hydroxypropyl methylcellulose			
	PEG–PCL–PEG	triamcinolone acetonide	Ocular inflammation	[48]
	Lecithin-based NPs embedded in poloxamers gel (P188 and P407)	dexamethasone		[49]
	PLGA–PEG NPs embedded in PEG–PLGA–PEG gel	triamcinolone acetonide	Age-related macular degeneration	[50]
	Bevacizumab-coated PLA NPs embedded in PLGA microparticles	bevacizumab		[51]
Dendrimeric nanocarriers	PEGylated polyamidoamine modified with cyclic arginine–glycine–aspartate hexapeptide and penetration	–	Posterior ocular diseases	[52]
	Timolol-derivatized polyamidoamine	timolol	Ocular hypertension	[53]
	Polyamidoamine/hyaluronic acid	antisense oligonucleotides	Regulation of the expression of target proteins and genes in cells	[54]

Table 1. *Cont.*

Type of Nanoparticle	Nanoparticle Composition	Drug Delivery	Treatment	Reference
Cyclodextrins	Propylamino-β-Cyclodextrin	latanoprost	Glaucoma	[60]
	γ-Cyclodextrin and randomly methylated β-cyclodextrin	celecoxib	Age-related macular degeneration and diabetic retinopathy	[61,62]
	α-Cyclodextrin/Soluplus/Pluronic P103	natamycin	Fungal keratitis	[63]
Polymeric vesicles	DOTAP/DOPE/DSPE–PEG	siRNA sequences/chlorhexidine	Keratitis caused by *Acanthamoeba*	[65]
	Precirol® ATO 5/castor oil/Span® 80/mPEG-2K-DSPE	natamycin	Fungal keratitis	[66]

PLGA—poly(lactic-co-glycolic acid); PVA—poly(vinyl alcohol); VEGF—Vascular Endothelial Growth Factor; PEI—poly (ethyleneimine); PVP—poly(vinylpyrrolidone); CH—chitosan; PCL—poly(ε-caprolactone); PEG—poly(ethylene glycol); DOTAP—1,2-dioleoylsn-glycero-3-trimethylammonium propane; DOPE—1,2-di-(9E-octadecenoyl)-sn-glycero-3-phosphoethanolamine; DSPE–PEG—1,2-distearoyl-sn-glycero-3-phosphoethanolamine-*N*-[methoxy(polyethylene glycol)-2000]; PEG-2K-DSP—*N*-(Carbonylmethoxypolyethylenglycol-2000)-1,2-distearoyl-sn-glycero-3-phosphoethanolamine.

3. Polymeric Nanoparticles in Cancer Diagnosis and Imaging

According to the WHO, cancer is the second leading cause of death worldwide, with an estimated 9.6 million deaths in 2018. These data indicate cancer to be one of the diseases with the highest rate of morbidity and mortality nowadays. Finding effective methodologies for early detection, diagnosis and treatment has become a fundamental objective when developing NPs as DDSs [67–69]. Ordinary imaging and diagnosis techniques can only detect tumor mass when it is at least one-centimeter in size, being notably difficult to detect cancer at early stages [70]. This is the reason many researchers are currently trying to develop new and smaller composites able to identify malignant cells related to cancer processes, in order to inform medical staff to devise a treatment strategy. Polymeric NPs have thus emerged as an alternative to limit ordinary contrast agents due to their surface modification abilities and their capacity to regulate solubility of the embedded agents in order to enhance imaging of cancerous cells. Some of the following recent investigations that have been consulted involve both therapeutic and diagnostic objectives (known as "theranostic agents"). This section of the review focuses on diagnostic and imaging results and just mentions some of the therapeutic facets.

3.1. Gold-Based Polymeric Nanoparticles Used in Cancer Diagnosis

Gold metallic NPs (AuNPs) and their derivatives are the most important investigation topic when describing new composites able to improve diagnosis and imaging techniques. Due to their versatility, they can be used in multiple imaging methods, providing high resolution and low or non-existent toxicity [71]. Computed tomography (CT) is one of the most commonly used diagnosis techniques in cancer imaging, mainly due to its low cost, high imaging resolution and compatibility with all types of patients. Scanning of soft tissues carried out by this technique requires contrast agents absorbing X-ray radiation. AuNPs have generated great interest as these agents, since they are nontoxic and present up to three-fold more efficiency in X-ray absorption than the current iodine-based CT contrasts agents. Other benefits related to AuNPs are the possibilities of designing and modifying their shape, size and surface. Although there are other NPs with a higher capability of X-ray radiation absorption, like bismuth-sulfide NPs, the control of their characteristics and the modification of their surface are more complicated [72,73]. In order to emphasize AuNPs contrast properties, encapsulation of these metallic NPs in polymeric NPs have been tested. Al Zaki et al. [74] designed and optimized polymeric micelles (AuMs) where 1.9-nm-size AuNPs were encapsulated within the hydrophobic core of micelles constructed from amphiphilic copolymer PEG–PCL. Blood pool contrast was obtained for 24 h and enhanced tumor margin delineation was observed, via CT, when AuMs were injected in living mice. Improvements in survival time when radiotherapy was applied were also demonstrated in these animals when treated with AuMs, compared to those which were not. Dedrimeric NPs were also investigated for the stable encapsulation of AuNPs for CT cancer diagnosis. Lin et al. [75] prepared a CD-derived

21-arm star-like triblock copolymer of β-CD-{PCL-poly(2-aminoethyl methacrylate)-poly[PEG methyl ether methacrylate]}. They combined a dendrimeric NP with the use of a CD unit in its nuclei, not only to stabilize AuNP as imaging agents, but also to embed doxorubicin to obtain a theranostic system. In vitro and in vivo experiments demonstrated the high-contrast properties of this system, characteristic of AuNP.

AuNPs can also be utilized in many other bioimaging techniques such as two-photon nonlinear microscopy, to study the binding coefficient between NPs and target cells and their absorption [76]. Single-photon excitation is a similar technique employed in vitro [77,78] to establish AuNPs accumulation in cells cytoplasm. Wang et al. [79] designed biodegradable polymeric NPs based on silica-coated AuNPs for photoacoustic imaging (PAI). This technique allows researchers to obtain images from biologic structures of different shapes and forms, even from organelles. It consists of the generation of wideband ultrasonic waves (called PA waves) due to thermoelastic expansion when a tissue is irradiated by near-infrared (NIR) light, which is absorbed by the target [80]. It is a very reliable technique to be linked to commonly used clinical diagnosing techniques. The gold nanospheres where synthesized, coated with silica, fluorinated and then introduced in a previously synthesized PLGA NP.

3.2. Gadolinium Polymeric Nanoparticles (GdNPs) Used in Cancer Diagnosis

Magnetic resonance imaging (MRI) allows three-dimensional high-resolution images to be obtained. It is useful for delimiting morphologic characteristics in tumors without producing ionizing radiation that could be harmful for the patient. This has become one of the best strategies in clinical cancer diagnosis [81]. To optimize this technique, contrast agents are utilized to enhance the variations between the different tissues, by lowering water relaxation parameter values (longitudinal or T1 and transverse or T2). There are many different types of contrast agents, but gadolinium-based materials are the most widely used [82] and mainly those formed by the chelated metal. While gadolinium-chelated complexes are easily eliminated from the organism by the kidneys because of their low molecular weight (<11 nm), if they are too big, they can be phagocytosed by macrophage cells (>200 nm) [83,84]. Nanotechnology has tried to overcome this inconvenience by designing new gadolinium-based contrast agents with enhanced imaging time, contrast effect and lowered toxicity, as well as granting passive targeting properties [85]. In order to modulate these characteristics, NP surface modification and full size control is necessary [86]. Some investigations have allowed enhancing of imaging by targeting key elements present in cancer cells, such as overexpressed surface proteins. To this end, Liu et al. [87] synthesized a novel multifunctional polymeric GdNPs-based contrast agent (Anti-VEGF PLA–PEG–PLL–GdNP). These nanoparticulate systems were designed with anti-VEGF antibody, which facilitates delivery to cancer cells in hepatocellular carcinoma (HCC) in order to improve its detection in early phases. Obtained NPs were 70–80 nm-sized, preventing them from being easily eliminated from the body. They managed to increase tumor area imaging time significantly in comparison with control substances. In a different approach, polymerization-induced self-assembly (PISA), a synthesis method used in organic chemistry, was applied by Esser et al. [88] to obtain polymeric NPs including Gd ions. The corresponding amphiphilic triblock copolymer poly(glycidyl methacrylate)–block–poly(oligoethylene glycol methyl methacrylate)–block–polystyrene was prepared, which, after self-assembly into the corresponding NPs, was further functionalized with Gd^{3+} chelates. Depending on the polymer composition (proportion of each block), the NP shape and size could be modulated. MRI contrast efficiency was also characterized, compared and classified in terms of size and shape, demonstrating that filomicelles were the most promising candidates as MRI contrast agents.

Gadolinium has also been used as an imaging platform in PAI technique. The great depth penetration that NIR light reaches, makes NIR-light-absorbing materials (650–900 nm) such as organic materials, the ideal candidates for this technique [89,90], even if they are optically unstable. Gadolinium-based agents could overcome this issue, enhancing both imaging time and resolution. Developing polymeric GdNPs where Gd-complexes can be attached and immobilized in macromolecules [91–93], red blood cells [94], monoclonal antibodies [95], etc., is a tedious and complicated process. Hu et al. [96] detailed a synthesis pathway to obtain a poly(isobutylene-*alt*-maleic anhydride) (PMA) framework pendent with perylene-3,4,9,10-tetracarboxylic diimide derivatives and PEG, able to self-assemble by ultrasound, to which Gd^{3+} are easily attached. The optimal characteristics of these systems for being used in living organisms was demonstrated when they were injected into mice; excellent biocompatibility and photostability, good water solubility, low toxicity, strong PA signal intensity and a good performance as contrast agents and their ability to passively accumulate in tumors by enhanced permeability and retention effect. Photothermal in vivo treatment improvement was also observed, due to strong NIR optical absorbance and perfect tumor ablation properties, along with the absence of apparent toxic side effects in normal tissues. Wu et al. [97] also described GdNPs specifically designed for MRI/CT/PAI guided photothermal therapy, whose composition was Gd–PEG-coated Bi. These NPs absorb NIR light and transform it into heat, increasing the temperature to 40 °C and producing the in vivo tumor ablation as well as its eradication.

3.3. *Perfluorocarbons Polymeric Nanoparticles (PFCNPs) Used in Cancer Diagnosis*

Perfluorocarbons (PFCs) are molecules whose structure is similar to common organic compounds (e.g., alkanes). The difference between PFCs and regular organic compounds is that every hydrogen atom is replaced by fluorine (^{19}F) in PFCs, the most electronegative element in the Periodic Table. This exchange grants new and interesting properties that can be useful for medical applications. Nuclear magnetic resonance (NMR) is usually based on the ^{1}H signal from the water of the body's tissues and mobile hydrocarbon compounds. There are also other nuclei such as ^{19}F [98,99], which can be used in this technique to improve diagnosis and imaging effects. Unlike hydrogen atoms, most of the fluorine found naturally in the organism are located in bone structures, which as solid structures, restrict fluorine signal for MRI assays [100]. One of the major problems connected with the use of PFCs, is their solubility, due to the fact that they have high hydrophobicity. Research and development of new systems able to load contrast agents and enhance their biodistribution, has led to the design of nanoparticulate systems which raise their imaging effects. Kristen Wek [101] designed and characterized a polymeric NP containing fluorine compounds for enhanced NMR effect and passive targeting using a described copolymer [102], obtained from polyethylene glycol methyl ether methacrylate (PEGMEMA) and trifluoroethyl methacrylate (TFEMA) monomers with an azide functional group. NPs were synthesized through atom transfer radical polymerization (ATRP) in order to obtain a small polydispersity index and provide precise molecular weights and sizes. This system also showed passive diffusion into tumors and irrelevant ^{19}F NMR signal alteration. In a similar approach, Pisani et al. [103] synthesized polymeric nanoparticles containing liquid PFCs which were sensitive to ultrasound imaging. They synthesized a single core of liquid PFCs and a homogeneous PLGA–PVA polymeric shell, in order to increase the solubility using a variation of the regular emulsion-evaporation methodology. Perfluorooctyl bromide (PFOB), perfluorodecalin (PFD), and perfluorohexane (PFH) polymeric NPs were successfully synthesized and PFOB nanomaterials were characterized and described as nontoxic. In a posterior research work, Giraudeau et al. [104] carried out further investigations and compared these PFOB NPs with free PFOB in several assays, obtaining promising results. Surface functionalization is also used when synthesizing these NPs in order to achieve higher performance levels. Diou et al. [105] added RGD (arginine-glycine-aspartic acid) peptide, commonly considered for active tumor targeting, to the surface of PEGylated polyester nanocapsules of PFOB by pre- and post-functionalization strategies. They were tested in vivo in mice bearing CT26 tumors by ^{19}F MRI, showing very interesting results.

3.4. *Other Nanoparticles Used in Cancer Diagnosis*

Although the above-mentioned NPs are the most common nanoparticulate systems currently used for cancer diagnosis, there are others which are under investigation. These are mainly based on different absorbing compounds or on a synergistic union among techniques and/or contrast agents in one single NP. Liopo et al. [106] described the synthesis pathway and characterization of PEGylated biodegradable melanin-like nanoparticles (MNP–PEG) and their properties used in photoacoustic tomography. MNP–PEG demonstrated biocompatibility with human MCF-7 and 3T3 cells and they remained stable in biologic medium for at least eight weeks. Belletti et al. [107] developed a synergy by joining two nanometric concepts, NPs and quantum dots (QDs). Their work was based on the concept that curcumin provokes apoptosis in primary effusion lymphoma (PEL) cells. However, this agent has a very low efficiency rate in this type of cancer treatment, due to its low solubility and consequently, low bioavailability. Encapsulating it in PLGA, NPs enhanced these characteristics and improved the amount of curcumin that was retained by the organism. QDs were also attached to the NPs surface as imaging agents, obtaining a theranostic application. The combination of two different metals in the particulate formulation was also explored. Another example was proposed by Zhou et al. [108] who combined Gd and Au advantages as contrast agents in MR and CT, respectively, to create new imaging agents for targeted dual mode tumor CT/MR imaging in vivo. In this research, PEI modified with folic acid and Gd chelators were used as a matrix to synthesize AuNPs. These systems were then complexed with Gd. Folic acid-targeted PEI-entrapped AuNPs loaded with Gd were characterized, showing good quality properties for in vivo applications: 3.0 nm size Au core, good water dispersion and nontoxicity. Regarding their imaging capabilities, a good X-ray absorption signal, higher than some other commercial contrast agents, and a reasonable r_1 relaxivity rate were shown, making them ideal candidates for dual mode nanoprobe use for targeted tumor CT/MR imaging in vivo. McQuade et al. [109] established a nanoplatform for theranostic purposes based on gold and superparamagnetic iron oxide NPs (SPIONPs) entrapped within a polymeric micelle, where amphiphilic diblock copolymer PEG–PCL acted as the polymer barrier in a similar assay as the aforementioned work by Al Zaki et al. [74]. On a higher level of complexity, Topete et al. [110] designed polymeric–gold nanohybrids to target multimodal theranostic agents, which are useful in optical and magnetic resonance. These folic acid-functionalized, doxorubicin/SPIONPs-loaded PLGA–Au porous shell NPs were tested in vitro in a human cervical cancer cell line in order to determine their physicochemical characteristics, cellular uptake and theranostic potential. They also reported an improvement in cellular uptake by applying an external magnetic field that guides the nanosystem to the cancer cells, as well as in targeting due to the folic acid.

Even though they are not considered strictly as polymeric nanoparticles, polymer-modified superparamagnetic iron oxide NPs (also called SPIONPs or Fe_3O_4 NPs) are remarkable contrast agents which are used in cancer imaging, thus deserving a special mention in this section [111–113]. Among their features, their capability to be surface-modified, in order to add polymeric agents that improve their imaging properties, as well as granting active targeting features, is a very powerful tool used in current investigations on developing new imaging agents [114–117].

4. Polymeric Nanoparticles in Oncologic Treatment

As previously mentioned, cancer has become a leading cause of death in developed countries. In fact, experts claim that over the next 20 years, the incidence of this disease is expected to increase by approximately 70% [118,119]. The classic therapeutic approach to deal with cancer consists of surgery, chemotherapy and radiotherapy. Chemotherapy is the treatment of choice in most cancers, but it does present high toxicity due to the affection of both healthy and cancerous cells [118–120]. Nanomedicine, defined as the use of materials in nanometric scale in medicine, offers a more specific alternative. Its main objective in oncology is to transport the drug only to cancer cells selectively in order to improve its efficacy and reduce its toxicity [118]. The potential application of nanomedicine can also provide early detection tools in cancer as well as combination therapies that can result in both better efficacy of treatment and prognosis [119].

4.1. Advantages of Nanotechnological Drug-Delivery Systems

Chemotherapeutic agents present several inconveniences, including poor aqueous solubility and nonspecific biodistribution. As a consequence, an inadequate drug concentration at tumors or cancerous cells, as well as toxicity to normal cells and possible development of multiple drug resistance, can frequently occur [121,122]. Nanomedicines have been shown to improve solubility of low soluble drugs and to reduce toxicity by dissolving them in their hydrophobic or hydrophilic compartment. Studies also report they have prolonged plasma half-life and a different biodistribution profile compared to conventional chemotherapy [121]. Their nanometric size, large surface-to-volume ratios and the ability for surface functionalization are crucial factors in their biodistribution in vivo [120]. The improvements noted in comparison to conventional free drug administration consist in improving the therapeutic index of the loaded chemotherapeutic active ingredients, increasing drug efficacy by achieving steady state therapeutic levels over an extended period of time, lowering drug toxicity due to controlled drug release systems and an improvement of solubility and stability [120,121]. Other notorious advantages of nanomedicines are the possibility of surface functionalization, as well as the possible combination of multiple drug-delivery systems to achieve a synergistic therapeutic effect. This field also allows the application of a combination therapy fusing chemotherapeutic and photothermal effects or creating magnetic nanostructures, making NP delivery easier with the application of an external magnetic field [121].

Currently, the nanoparticulate systems approved for their clinical utilization are liposomes, albumin-based NPs, polymeric NPs and inorganic NPs [120]. Some polymer-derived liposomes or even polymer–inorganic NP composites, can be found in literature. This section will focus on the polymeric NPs as DDSs in oncology. In the last decade, biodegradable polymeric NPs have been widely considered as potential drug-delivery systems for their application in oncology. Polymers are the most common materials for constructing NP-based drug carriers. Due to their nature, they make possible the customization of many properties, such as hydrophobicity, molecular weight and biodegradability [119]. In most cases, polymeric NPs are spherical and are constituted of dense matrices. PLA, polyglycolic acid (PGA) and PLGA are the most common synthetic polymers used in these carriers. Natural polymers, such as dextran, gelatin, guar gum, collagen and CH can also be utilized. All of them are completely biocompatible and biodegradable, and have acknowledged degradation curves, making the drug release process of these nanocarriers more accessible to be customized in comparison to other nanoparticulate DDSs [118,119,123]. Moreover, performing nanoparticle-surface modifications by using different types of ligands to the receptors over expressed on the cancer cells, these polymeric NPs can be targeted directly to them [121].

Currently, albumin is being widely used for the preparation of NPs as DDS due to its intrinsic characteristics. It is a very stable, soluble, acidic, non-immunogenic, nontoxic and biodegradable protein. In addition, it has a high conjugation capacity due to its several binding sites and a long half-life (19 days). It accumulates naturally in the tumor environment due to its advantageous incorporation. When albumin interacts with some epithelial cell surface receptors, it is actively transported via transcytosis, a process that includes both endocytosis and exocytosis. One of the most relevant receptors, except in the brain, is glycoprotein 60 (Gp60), which is overexpressed in vascular endothelium cells that compose tumor blood vessels. Along with this receptor, the accumulation of albumin is even more facilitated by SPARC glycoprotein [124]. This protein has a significant homology to Gp60, and its overexpression is related to cancerous processes and the higher rate of binding and subsequent uptake by tumor cells has been described in several tumor model experiments. These facts, united to the enhanced permeability and retention effect (EPR) in solid tumors, which is discussed in detail in this work, allows albumin-based NPs to be directed straight to tumoral cells without the need of active targeting with antibodies or other mechanisms [125–128].

4.2. Challenges Associated with Nanoparticulate Drug-Delivery Systems

Despite their advantages and formidable potential, some inconveniences in their use can be found related to their limited shape, chemistry, wide size distribution agglomeration state and electromagnetic properties that can lead to poor oral bioavailability, instability in circulation and inadequate tissue distribution. The vast majority of polymeric NPs are spherical, yet a wide range of different sizes may be generated during synthesis. Their continual interaction with living cells could trigger a range of adverse effects in humans and animals [121,122]. To resolve these problems, new techniques are under investigation, being particle replication in no wetting templates (PRINT) the most recent approach in this field. PRINT technique concedes the synthesis of uniform polymeric NPs, permitting the customization and stabilization of properties such as shape and size. Therefore, the amount, rate and pathway for the uptake of the encapsulated drug in the formulation could be tailored as well [119].

4.3. The Enhanced Permeability and Retention (EPR) Effect

The EPR effect is a particular paradox that occurs only in solid tumors, and it is directly related to their pathophysiological and anatomic characteristics, which differ from normal tissues. This effect was first described by Matsumura and Meda in 1986 and has been studied since then. Most solid tumors have abundant, but aberrant, vasculature and poor lymphatic drainage because of the compression of the cancerous cells and the dysfunctional lymphatic angiogenesis [120,129–131]. Their blood vessels have an abnormal architecture and produce an excess of a variety of vascular permeability factors, resulting in an increased vascular permeability to ensure the supply of oxygen and nutrients to tumor tissues [129,130]. In addition, it has been demonstrated that the endothelial junctions of tumor blood vessels are larger than normal. Thus, macromolecules larger than 40 KDa, can percolate through these enlarged junctions and accumulate in the interstitial fluid and environment of the tumor. Therefore, the EPR effect depends on the molecular weight of molecules or particles, being only those larger than 40 KDa, which correspond to the limit of renal clearance, the ones that experiment this effect (Figure 3). In addition, these macrostructures remain in the tumor for a considerably long time (several days). The EPR effect has become the principal mechanism to consider in oncologic treatment design. Nanocarriers are designed to take advantage of this EPR effect and accumulate in the tumor environment to achieve the best targeting and therapeutic efficacy [120,131]. Only poorly vascularized tumors, mainly pancreatic, prostatic and liver metastatic experience less EPR effect than other types of cancer. Meda et al. [132] discovered that substances that emulate the effect of vascular mediators involved in EPR effect could enhance it, principally nitric oxide (NO), as well as bradykinin, prostaglandins and VEGF, by facilitating tumor angiogenesis and growth. Recent developments have also demonstrated that another vascular mediator, carbon monoxide (CO), also enhances this EPR effect. It has been demonstrated that this effect was more present when the systolic blood pressure was higher than normal. In order to intensify the EPR effect, blood pressure can be elevated, or NO-releasing and CO-releasing agents can be administered [129,130,132,133]. It is important to mention that, with elevated blood pressure, studies showed that the accumulation of nanocarriers in the tumor was significantly higher and the release of these substances in healthy tissues was lower, due to vasoconstriction and compaction of endothelial junctions, translating in less toxicity [129].

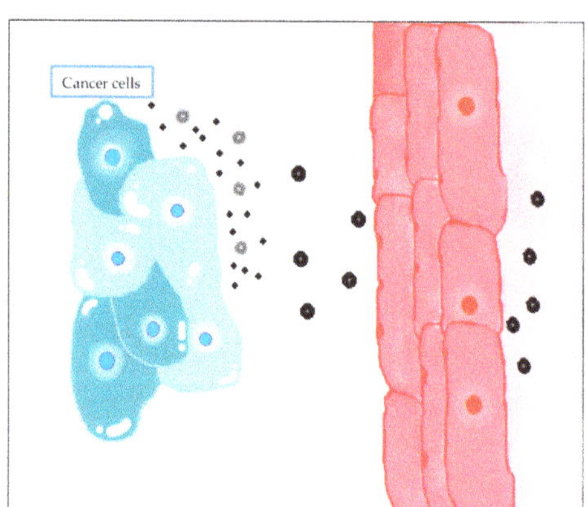

Figure 3. Schematic of the EPR effect: NPs pass through the endothelial fenestrations and reach cancer cells.

This pathway could indeed be beneficial in order to deliver NPs to the tumor, but it also encounters several challenges. First, an increment of interstitial fluid pressure due to a leak of proteins caused by the tumor growth, supposes a barrier that can block the penetration of NPs inside the tissues. The pressure on the abnormal lymphatic vessels caused by the tumor cells, provokes a considerable reduction in their drainage, contributing to the accumulation and increment of the fluid pressure. Finally, the heterogeneity of the tumor tissues leads to an abnormal distribution of the nanocarriers, as the central part of the tumor is normally less vascularized or necrotic. Indeed, penetration of the NPs to this central area is poor via EPR [120,122,131].

4.4. Active Targeting

This method, based on molecular recognition processes, consists of modifying the NPs surface with one or more required moieties to achieve their functionalization and, in consequence, raising the drug concentration in tumor tissues. The most frequent targeting components are monoclonal antibodies and antibody fragments, antigen biding such as fragments and single chain variable fragments (Figure 4). Other molecules that can be utilized are transferrin, enzymes, folic acid and other proteins or peptides. These compounds recognize and bind receptors in the cancerous cells. Normally these ligands are selected depending on which receptor is more overexpressed in tumor cells compared to normal tissues [118,120]. If the selected receptors are internalizing, then the nanocarriers will be transported into the cancerous cells by a specific pathway. In addition, these ligands can be directed to endothelial cells of solid tumors blood vessels to improve the accumulation of nanocarriers in the site of action [134]. Therefore, the purpose of active targeting is to improve the incorporation of the nanocarriers to the cancer cells. Active targeting is a complementary approach to the EPR effect, to improve addressing to the tumor.

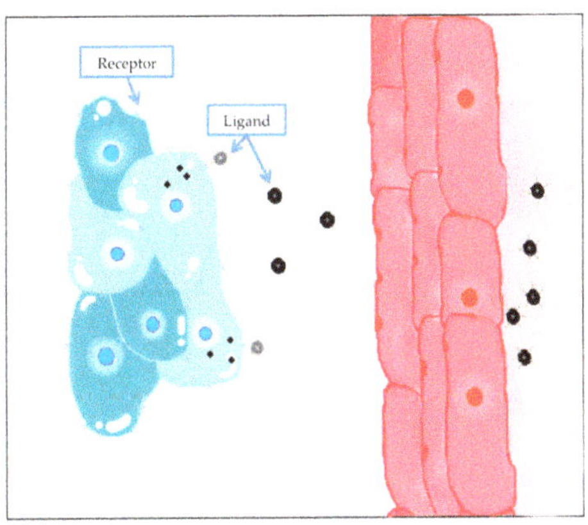

Figure 4. Schematic of the active targeting process: NPs have been functionalized by adding ligands onto their surface that can recognize and bind the receptors in cancer cells.

Some of the most relevant receptors that are overexpressed in cancer cells are adenosine, transferrin, somatostatin and folate receptors, epidermal growth factor receptor (EGFR), glucose, integrins, chlorotoxin and cytokeratin [135].

Finally, the election of the ligand depends on numerous factors, such as the type of targeted tumor, characteristics of the overexpressed receptors, if internalization process is possible and the proper structure and biodistribution of the nanocarrier [134].

4.5. *Stimuli-Responsive and Triggered Release Systems*

The objective of these systems is the controlled release of antineoplastic drugs provoked by stimuli that develop a change in the nanocarrier (Figure 5). Both internal (changes in pH, redox, ionic strength) and external stimuli (temperature, magnetic fields or light) can trigger the release of drugs [119].

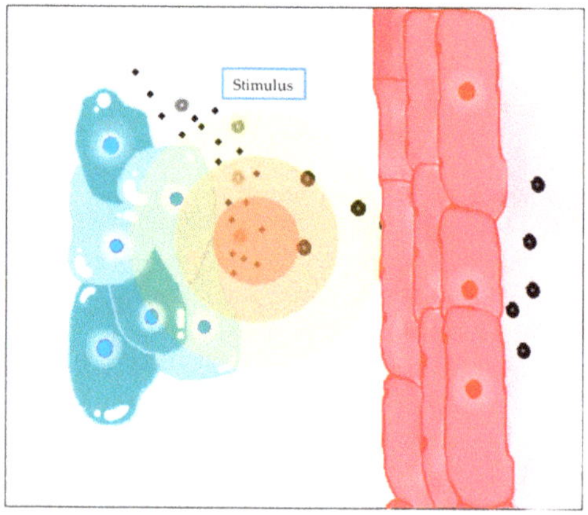

Figure 5. The exposure to a stimulus triggers congregated nanoparticles drug release.

To date, many polymeric NPs have been used in the carriage of antineoplastic drugs like paclitaxel, doxorubicin or camptothecin in many types of cancer. The use of these polymeric NPs can provide improvements in cancer therapy by exploring new routes of administration of some drugs, combining some active substances to potentiate their action or combining with other therapies like gene therapy. Ahmad et al. [136] proposed an enhancement of doxorubicin oral bioavailability through surface modified biodegradable polymeric NPs as an alternative to intravenous administration. They studied drug-loaded PEGylated PLGA NPs pharmacokinetics compared to doxorubicin in Wistar rats. Results showed that NPs had better activities and also higher bioavailability compared to oral drugs. Soma et al. [137] studied the synergistic effect of polyalkylcyanoacrylate (PACA) nanoparticulate formulation of cyclosporin A and doxorubicin compared to only NPs in resistant tumors. Results showed that the combination of both active ingredients were more effective in terms of growth rate inhibition in P388/ADR cells.

There have been numerous examples of the use of polymeric NPs in concrete oncologic diseases. Albumin-bound (nab)-paclitaxel NPs (Abraxane®) were approved in 2012 by the US Food and Drug Administration (FDA) for cancer treatment. It has been used since then for the treatment of a large list of cancer including non-small cell lung carcinoma, metastatic breast cancer and pancreatic cancer. These NPs were developed to improve the pharmacokinetics and pharmacodynamics of paclitaxel and also to avoid the toxicities of polyoxyethylated castor oil solvent (Cremophor), used for paclitaxel because of its poor aqueous solubility. In addition, these NPs, in combination with gentamicin, have slightly improved survival rate in advanced and metastatic pancreatic cancer. One of the latest nanoformulation of novel paclitaxel liposome–albumin composite obtained a high encapsulation efficiency of 99.8% [138].

In brain tumors, Cırpanlı et al. [139] studied the activity of camptothecin-loaded cyclodextrin NPs in a rat glioma model. This nanoparticle suspension was injected by convectional enhanced delivery at the same coordinates where the tumor cells were. The use of nanomaterials prevented the drug from hydrolysis and allowed its action. Results showed an improvement of the survival time and determined camptothecin-loaded amphiphilic cyclodextrin nanosystems as an effective nanocarrier. Concurrently, placlitaxel-containing PEG–PLGA NPs coated with AS1411, a DNA aptamer which binds to a protein highly expressed in the surface of cancer and endothelial cells of gliomas, were developed by Guo et al. [140]. Results in vitro and in vivo showed higher tumor growth inhibition compared to placlitaxel-NPs alone and Taxol®. In vivo experiments were carried out in Sprague–Dawley (SD) rats, Wistar rats and nude mice and the formulation was administered via the tail vein. Malinovskaya et al. [141] studied the improvement of crossing the blood–brain barrier, which is the principal issue in the therapy of intracranial tumors, using PLGA NPs overcoated with poloxamer 188 for the treatment of glioblastoma in U87 human cells. Hekmatara et al. [142] investigated a system constituted by doxorubicin bound to polysorbate 80, that was in turn coating poly(butyl cyanoacrylate) NPs, in a orthotopic rat 101/8 glioblastoma model, in comparison to doxorubicin in solution, both administered via intravenous injection. The group treated with this nanosystem showed better antitumor effect compared to the control.

Breast cancer is the most prevalent cancer in women, representing an overwhelming 30% of all diagnosed cases of this disease. The notorious diversity in subtypes of breast cancer and their variable response to distinct treatments lead to a great difficulty to develop a universally effective treatment [143]. The use of nanocarriers in the treatment of this type of cancer opens a door to improve their effectiveness. Yuan et al. [144] studied the action of pH-sensitive PEG–PLGA–PGlu (polyglutamic acid) NP embedded with curcumin and doxorubicin in drug resistant cancer stem cells and tumor cells of breast tumors, obtaining good results in mice models. The combination of the use of nanoparticulate systems with photodynamic therapy in breast cancer were investigated by Hu et al. [145]. They developed oxygen-generating theranostic NPs of poly(ε-caprolactone-co-lactide)-*b*-PEG-*b*-poly(ε-caprolactone-co-lactide) with doxorubicin, chlorin e6 and colloidal MnO_2 to generate oxygen in the tumor environment, relieving tumor hypoxia and improving photodynamic therapy and doxorubicin action. MDR has also been investigated for breast cancer. Shafiei-Iranneja et al. [146] demonstrated that the use of polymeric NPs is a good strategy to combat MDR in doxorubicin-resistant breast cancer (MCF-7/DOX) cells. However, the nanoencapsulation of these NPs together with D-α-tocopheryl polyethylene glycol 1000 succinate (TPGS), a compound used for surface modification of PLGA NPs, has shown a higher cytotoxicity and apoptosis in breast cancer cells [147]. In addition, a higher intracellular drug accumulation and a reduced drug efflux, associated with a decreasing cellular ATP content and an inhibition of *P*-gp activity, have been observed [146]. Additionally, Vitamin E Succinate (VES) that exists as the hydrophobic moiety of TPGS, has also been shown to represent a promising strategy for delivery of doxorubicin into MCF-7/ADR cancerous cells and to revert MDR [148]. Treatment options for triple-negative breast cancer subtype are narrowed down to traditional chemotherapy, surgery and radiation. It is well known that these treatments are not tumor selective and are not very effective, especially when metastatic disease is present. Khanna et al. [143] proposed the use of perlecan-targeted PLA–PEG–maleimide NPs for drug delivery in the treatment of this specific type of breast cancer. The overexpression of a cell surface protein, perlecan (HSPG2), has been recently identified, in this particular type of cancer. This molecule is a large basement membrane protein that is remarkably glycosylated, and it plays a role in binding growth factors. Regarding this fact, researchers have developed two monoclonal antibodies (Clone 6 and AM6) that attach with great affinity to perlecan (HSPG2) present in tumor cells. Indeed, paclitaxel-loaded PLGA NPs were functionalized with these two antibodies using thiol–maleimide chemistry. The antibodies were covalently conjugated to NPs without affecting antibody binding affinity or NP properties. Results of in vitro and in vivo models of triple-negative breast cancer showed that perlecan-targeted NPs improved cell uptake, retention, cytotoxicity in vitro and enhanced tumor growth inhibition in vivo.

Liver cancer has become one of the most frequent cancers and its mortality rate is considerably high, being the third cause of death provoked by oncologic pathologies. Most antineoplastic drugs have high liver toxicity and can trigger severe side effects. Polymeric NPs have been employed as promising carriers for anticancer drugs, not only to improve their efficacy, but also to reduce the appearance of side effects. Zhu et al. [149] synthesized a nanosystem based on a new galactosamine conjugated polydopamine-modified copolymer (Gal–pD–TPGS–PLA). In vitro cellular uptake and cytotoxicity assay showed that Gal–pD–TPGS–PLA NPs target HepG2 cells via ASGP receptor-mediated recognition and remarkably inhibit cell proliferation. In addition, docetaxel-loaded Gal–pD–TPGS–PLA NPs reduced tumor size more, evidently in vivo, than Taxotere®, docetaxel-loaded TPGS-PLA NPs or pD-TPGS-PLA NPs or saline.

Table 2 shows a summary of the nanoparticulate systems as DDS for cancer treatment presented in this review.

Table 2. Anticancer drug polymer nanoparticles organized by cancer type they are used for.

Polymer	Active Principle	Type of Cancer	Experimental Model/Route	Size (nm)	Z Potential (mV)	PDI	References
PEGylated PLGA	doxorubicin	various	In vivo: Bioavailability assay in Wistar rat Oral	183.10	−13.10	0.132	[136]
PACA	doxorubicin–cyclosporin A.	various	In vitro: P388/ADR cells line	288	*	*	[137]
Lip–BSA	paclitaxel	various	In vivo: 4T1 cells in BALB/c mice Tail vein	116.2	−18.4	0.307	[138]
PCL–PEG	camptothecin	glioma	In vivo: 4T1 cells in BALB/c mice Tail vein	274	−19	0.07	[139]
PLGA–PEG	paclitaxel	glioma	In vivo: gliosarcoma 9L cells in Fischer F344 rats Direct injection	121	23.7	0.088	[140]
PLGA–Cyanine5.5	doxorubicin	glioblastoma	In vivo: C6 Glioma cells in Wistar rats and nude mice Tail vein	114	−14.9	0.196	[141]
PBCA	doxorubicin	glioblastoma	In vitro: U87 glioblastoma human cells line	260	−19	0.02	[142]
PLA–PEG–maleimide	paclitaxel	breast cancer (TNB)	In vitro: MDA-MB-231 cells In vivo: BALB/c homozygous nude mice Intravenous injection. Tail vein	212	−16.34	0.183	[143]
mPEG–PLGA–PGlu	doxorubicin–curcumin	breast cancer	In vivo: LM2 cells in BALB/c homozygous nude mice Tail vein	107.5	−13.7	*	[144]
PCLLA–PEG–PCLLA	doxorubicin and Chlorin e6-MnO_2	breast cancer	In vivo: MCF-7/ADR cells xenograft in female BALB/c nude mice. Tail vein	120	−8.9	*	[145]
TPGS–PLGA	doxorubicin and metformin	breast cancer	In vivo: MCF-7 cells in nude mice Tail vein	87	−3.5	0.5	[146]
TPGS–PLGA	docetaxel and salinomycin	breast cancer	In vitro: MCF-7/DOX cell line	73.83	−25.7	0.193	[147]
Gal–pD–TPGS–PLA	docetaxel	liver cancer	In vivo: MCF-7 cells in BALB/c mice Orthotopic injection	209.4	13.7	0.145	[149]
PCL–PEGPEG–PCL	paclitaxel	lung cancer	In vivo: MCF-7/ADR cells in BALB/c nude mice Intravenous injection	168	−12.49	0.19	[150]
PEI–PLA	paclitaxel	lung cancer	In vivo: A549 cells in BALB/c mice. Tail vein	67.31	30.3	0.105	[151]

PCLLA—poly(caprolactone-co-lactide); PBCA—poly(butylcyanoacrylate). * Data not included in the corresponding publication.

Lung cancer is also one of the most prevalent. Hu et al. [150] explored the efficacy of paclitaxel-loaded polymeric NPs combined with circadian chronomodulated chemotherapy. In vitro results showed that this nanosystem exhibits best anti-cancerous activity against A549 cancer lung cells compared to paclitaxel and also determined that the best time of the day to be administered was 15 h after sunrise. In addition, the use of smart PEG-derived polymeric NPs to codeliver paclitaxel and siRNA against survivin gene in lung cancer was proposed by Jin et al. [151]. In vitro results demonstrated that the nanoparticulate formulation presented less toxicity and more antiproliferation effect of paclitaxel on A549 cancer lung cells. In addition, in vivo studies showed accumulation of NPs in the tumor environment and their ability to impede tumor growth. The survival rate was higher because of the-silencing of surviving gene and the action of paclitaxel into tumor cells. The use of inhalable nanocarriers to deliver antineoplastic drugs has been developed recently due to their specific characteristics. These nanocarriers highly associate with drugs and they sustain their release. They can be efficiently transferred into aerosols and remain highly in nebulization state. They also have the capacity to avoid mucociliary clearance as well as respiratory phagocytic mechanisms, thus prolonging the permanence of the antineoplastic drug within the respiratory tract. Polymeric NPs have been extensively used for the aerosol delivery of chemotherapeutics, genes or their combination for lung cancer therapy. Different examples of tested nanoparticulate systems prepared as inhalable polymeric nanocarriers are enumerated in Table 3.

Table 3. Inhalable polymeric nanocarriers.

Polymeric Nanocarrier	Active Principle	Preparation Method	Size (nm)	Z Potential (mV)	PDI	References
Gelatin NPs	cisplatin	desolvation	220	−9.3	0.287	[152–154]
HSA NPs	doxorubicin + TRAIL	self-assembly	341.6	*	*	[155]
CH/PLGA NPs	OMR	emulsion–diffusion	160	29	0.033	[156,157]
BIPCA NPs	doxorubicin	emulsion polymerization	137.2	23.5	0.12	[158]
PEGylated PAMAM dendrimers	doxorubicin	chemical conjugation	26.1	−6.6	0.108	[159]
Hyaluronan conjugates	cisplatin	covalent bonding	*	*	*	[160]
PEI polyplexes	p53	electrostatic complexation	*	*	*	[161]
SDA–PEI polyplexes	PDCD4 + shAkt1	electrostatic complexation	*	*	*	[162]
Glucosylated PEI polyplexes	PTEN	electrostatic complexation	*	*	*	[163]
UACH polyplexes	PDCD4 PTEN	electrostatic complexation	*	*	*	[164,165]
SPE–GPT polyplexes	shAkt1	electrostatic complexation	163.2	9.14	0.192	[166]
SPE–PEG polyplexes	PDCD4	electrostatic complexation	130	8.61	1.13	[167]
CH–g–PEI polyplexes	shAkt1	electrostatic complexation	166.4	−20	*	[168,169]
PLL/protamine polyplexes	p53	electrostatic complexation	*	*	*	[170]
PEI–alt–PEG polyplexes	Akt1 siRNA	electrostatic complexation	*	*	*	[171]
PEI polyplexes	doxorubicin + Bcl2 siRNA	electrostatic complexation	78.2	20.4	*	[172,173]

HSA—human serum albumin; TRAIL—tumor necrosis factor-related apoptosis-inducing ligand; OMR—antisense oligonucleotide 2′-O-methyl-RNA; BIPCA—poly(isobutyl cyanoacrylate); PAMAM—poly(amidoamine); p53—tumor suppressor gene; SDA—sorbitol diacrylate; PDCD4—cDNA of programmed cell death protein 4; PTEN—phosphatase and tensin homolog deleted on chromosome 10 gene; UACH—urocanic acid–modified chitosan; SPE–PEG—spermine-*alt*-poly(ethylene glycol) polyspermine; shAkt1—hRNA-silencing Akt1. * Data not included in the corresponding publication.

The main inconvenience related to the use of inhalable NPs is that—due to their nano-range size (<0.5 μm)—they could be quickly exhaled even before they reach the site of action. Currently, other methods to improve pulmonary drug delivery are being investigated. Principally, the combination of both microparticles (MPs) and NPs are the main option. MPs permit deeper lung deposition, but they are opsonized by alveolar macrophages. Studies are directed to develop a functional combination of NPs and MPs, where the MPs act as transporters to deeper areas in the lung where NPs could not reach [140]. Although these protocols have an enormous potential, they are not exempted of health risks. First, the deposition of insoluble nano- or micro-carriers can provoke a local inflammatory response and oxidative stress. It has been demonstrated that surface charge of NPs has a significant role in lung toxicity. While anionic biodegradable NPs showed good tolerability, cationic NPs caused toxic effects. On the other hand, NPs can react with the pulmonary surfactant and provoke a quick decrease in the surface tension during compression/expansion cycles, resulting in life-threatening consequences to the patient. The development of safer nanocarriers for pulmonary drug delivery is being intensively researched. Some physicochemical properties are related to the harmful potential, such as particle size, shape, structure, biodegradability and surface charge. Active targeting via surface modification is also being investigated, to enhance NPs accumulation into lung cancer cells via receptor-mediated endocytosis.

The use of carriers for the intracellular delivery of macromolecules of DNA or siRNA for nanoparticle-based gene anticancer therapy is considered the next generation of medicine. Polymeric nanoparticles are one of these nanocarriers. In fact, they can be used in combination with other therapies, for example, the codelivery of chemotherapeutics drugs and small siRNA by using smart polymeric NPs with pH-responsive and PEG-detachable properties, was explored by Jin et al. [151] with promising results. In cancer therapy, the goals of the transmission of genetic material into specific cells can be the correction of the mutation present, RNA interference, trigger the immune response against cancer cells, induce an antiangiogenesis effect, produce cytotoxic proteins or produce enzymes that helps in the activation of some antineoplastic drug. The nanocarriers must be capable to deal with numerous obstacles related to the internalization of the genetic material such as cellular membranes (extra and intracellular), the process of endocytosis and the later breakout from endosome and nucleus. In addition, they must be biocompatible, nontoxic, non-immunogenic and stable and must be able to protect the genetic material from degradation, characteristics that are applicable to polymeric nanoparticles. In the last years, this field has advanced notoriously and has been directed towards the development of multifunctional NPs for cancer treatment and diagnosis. At the moment, the codelivery of chemotherapeutic drugs and genetic material with a synergistic effect is the main subject of study. It is important to highlight that the risks associated with the use of these therapies are not well known and the results obtained so far are not conclusive. In fact, some studies suggest that NPs could interfere with cellular biologic functions, including at genetic levels. Because of that, nanotoxicology is also a developing field, very necessary to a future establishment of regulations and guidelines [174]

5. Polymeric Nanoparticles as Nutraceutical Agents

Although there is no official accepted definition of nutraceuticals, they are mostly referred to as pharma-foods, a powerful toolbox to be used as a complement to the diet and before prescribing drugs, in order to improve health and prevent and/or treat pathologic conditions. Subjects could be people who may not yet be eligible for conventional pharmaceutical therapy [175]. There is widespread inconsistency and confusion in the definition of "nutraceuticals". Substances from similar sources are classified differently, such as plant-derived drugs, for example, digoxin from foxglove leaves is in the group of the medicinal products, while extracts from green tea leaves are regarded as nutraceuticals [176] (Figure 6). Regarding the legislation, in the United States of America, the FDA regulates dietary supplements, which include nutraceuticals, under the Dietary Supplement Health and Education Act of 1994 (DSHEA). In contrast to Canadian regulations, research studies in humans to prove dietary supplement safety/efficacy, are not required by the FDA prior to marketing [177]. The current European regulations consider nutraceuticals as belonging to the same category as food supplements. The Directive 2002/46/EC on food supplements and novel foods, which was recently modified by the new European Parliament and Council Regulation (EU) 2015/2283, defining new foods categories, completes the classification of food supplements, but it still does not mention the term 'nutraceutical' [178].

The use of nutraceuticals for several pathologies has been reported. Some nutraceuticals can be used to reduce some of the main cardiovascular risk factors, such as altered blood glucose levels, hypertension and hypercholesterolemia [179]. The most frequently occurring cholesterol-lowering and blood–pressure lowering substances found in nutraceuticals are the following: berberine, beta-glucans, sterols, isoflavones, mono unsaturated fatty acids and monacolin K (also known as lovastatin) from extracts of red yeast rice fermented by *Monascus purpureus* [178,180] or the use of potassium, magnesium, L-arginine, vitamin C, cocoa flavonoids, beetroot juice, coenzyme Q10, melatonin and aged garlic extract [181]. In the case of the glucose metabolism and type 2 diabetes mellitus (T2D) the evidence suggests that increasing omega-3, omega-6 or total polyunsaturated fatty acids (PUFAs) has little or no effect on prevention and treatment of T2D [182], but randomized controlled trials suggest that viscous dietary fiber at a median dose of ~13.1 g/day may offer beneficial effects on glycemic control and, thus, an improved cardiovascular disease risk profile [183]. In addition, vitamins, mainly vitamin C and vitamin D, have been recommended as nutraceuticals to reduce periodontal risks or improve periodontal health [184]. Riboflavin, coenzyme Q10, magnesium, butterbur, feverfew, and ω-3 PUFAs have been recommended for adults with migraine [177]. Nevertheless, the evidence of the efficacy of nutraceuticals for the treatment of pediatric migraine is limited [185]. Many nutraceuticals have been considered useful, not only to treat some pathologies, but also to mitigate disease-related symptoms. In osteoarthritis, a chronic disease, the nutraceuticals may represent promising alternatives for the relief of pain, where the conventional pharmacological approaches to pain relief and joint repair have not always been safe for long term use [186].

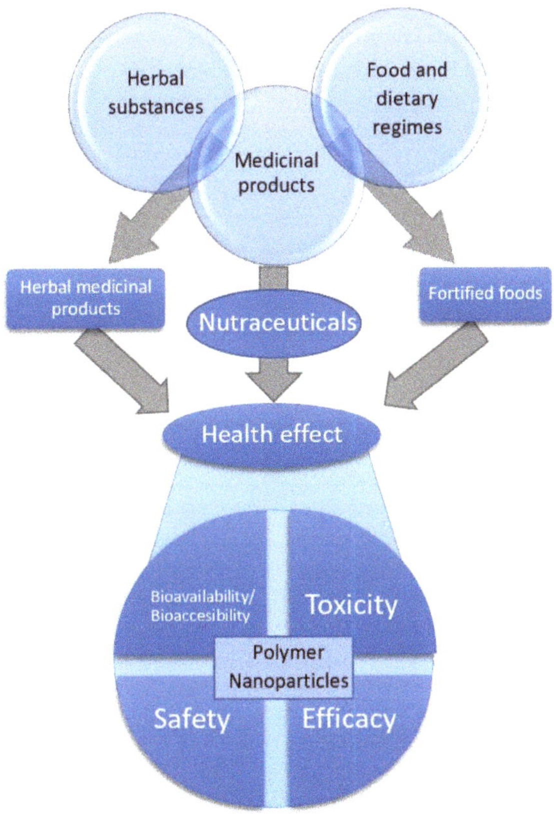

Figure 6. Use of polymeric nanoparticles for nutraceuticals and different bioactive compounds for greater health and medical benefits.

5.1. Bioavailability and Nanoparticles

As previously mentioned, the bioavailability refers to the portion of the bioactive compound that is absorbed in the body entering systemic circulation and performing functions. In order to determine bioavailability, it is necessary to measure blood plasma levels [187]. There are many animal model and epidemiological studies associated with food supplements or nutraceuticals, indicating their effectiveness and safety, however, the bioavailability is not clear [188–190]. The bioavailability of several nutraceuticals depends on many factors, such as dosage, possible interaction with the food matrix, like protein and fibers, the hydrophobicity of the compound, low chemical stability, intestinal first-pass metabolism [178,187,191] and gut microbiota, which can catabolize non-absorbed nutraceuticals and generate metabolic products that can have physiologic effects, and at the same time, prebiotic properties [192]. For example, the bioavailability of quercetin, (a flavonoid, known for its vascular function), is conjugated into glucuronide/sulfate metabolites, before being absorbed, reaching approximately μM levels within a few hours after intake and a half-life of around four hours [193]. quercetin presents a low bioavailability, probably attributed to its poor affinity with the different lipid phases found in the small intestine, inhibiting the uptake into CaCo-2 cells (intestine cells) [194]. However, the quercetin prenylation to 8-prenyl quercetin (8-PQ) is used as a strategy for elevating its lipophilicity and exert anti-inflammatory effects stronger than non-prenylated quercetin in macrophage cells [195].

Recently, new DDSs on the order of nanometers, in the nanometer range, are being engineered to improve the solubility of hydrophobic compounds, minimize systemic side effects and/or enhance the bioaccessibility and bioavailability of nutraceuticals (Table 4). Bioaccessibility is a property that refers to the quantity of a compound that is released from its matrix in the gastrointestinal tract, becoming available for absorption and reaching blood stream. Nanometric delivery systems, derived from food-grade phospholipids and biopolymers, adopt many forms, including liposomes, micelles, micro/nanoemulsions, NPs, polyelectrolyte complexes and hydrogels. The small particle size and the customized materials used to create delivery systems offer some unique properties, such as higher bioaccessibility, stability and resistance to enzymatic activity in the gastrointestinal tract [196]. The polymer NPs, typically assembled from dense proteins and polysaccharides matrix, increase the bioavailability and bioaccessibility of bioactive compounds, due to higher water solubility, with a tendency to increase when mixed with oil droplets. This takes place by promoting solubilization of the bioactive in the micelle phase of the small intestine. Its bioaccessibility depends on bioactive-polymer interactions and susceptibility to digestive enzymes [187]. For example, resveratrol presents a high intestinal absorption (>70% of the administered dose), but a low oral bioavailability (less than 1–2% of the dose). Calvo-Castro et al. [197] showed a new approach to significantly increase the hydrophilicity and thus, the bioavailability of resveratrol, using a liquid micellar formulation, without any adverse effects. It has been reported that resveratrol in ovalbumin–carboxymethyl cellulose NPs improves the photostability of trans-resveratrol when it is exposed to UV light and releases profile in the in vitro simulated gastrointestinal tract [198]. The development of a novel self-microemulsifying formulation (Capryol 90, Cremophor EL and Labrasol) for codelivery of resveratrol and curcumin, one of the most described nutraceuticals, results in an enhanced oral absorption and an improvement of the poor oral bioavailability of both compounds, which are not very water-soluble [199]. In addition, the interest of curcumin has increased because of its synergistic effects in addition to conventional therapeutic agents for various diseases, especially cancer [200,201]. In ovarian cancer therapy, the toxicity caused by triptolide, a potential anticancer agent, may be reduced by curcumin, due to its antioxidative stress through mPEG-DPPE/calcium phosphate NPs [202].

Table 4. Development and use of nutraceutical-loaded polymeric nanoparticles on different pathologies.

Drug Delivery	Polymeric Nanoparticle	Experimental Model/Route	Results	Reference
CD98 siRNA plus curcumin	HA-functionalized NP encapsulated in hydrogel (CH: alginate; 3:7)	In vitro: Caco2-BBE and Raw 2647 cells	↑Cellular uptake ↓Expressions of CD98 and TNF-α	[203]
		In vivo: DSS-induced UC/orally	↓Weight loss ↓Fecal Lcn-2 levels ↓MPO activity ↓Histological damage ↓CD98 and TNF-α mRNA expression	
curcumin plus celecoxib	pH sensitive enteric polymer NP (Eudragit® S100)	In vivo: TNBS-induced UC/orally	↓MPO, SOD and LPO ↓Leukocyte infiltration	[204]
curcumin	Biopolymeric CH NP	In vitro: HeLa cells	↓Proliferation and viability cell ↑Apoptotic activity, DNA damage, cell-cycle blockage and ROS levels	[200]
curcumin	PLGA NP	In vivo: MIA-induced OA/orally	↑Cellularity and matrix	[205]
curcumin	Theracurmin® NP	In vivo: DSS-induced UC/orally	↓NF-κB, TNF-α, IL-1β, IL-6, CXCL1 and CXCL2 and neutrophil infiltration ↑CD4+ and Foxp3+ T cells ↑CD103+ and CD8α− dendritic cells ↑*Clostridium* cluster IV and XIVa ↑Butyrate levels (bacteria and fecal)	[206]
resveratrol plus quercetin	PEG modified CH NP	Ex-vivo: Albino rabbit cornea	↑Solubility and permeation ↓Intraocular pressure	[207]
docetaxel plus resveratrol	EGF conjugated core-shell lipid–polymer hybrid NP	In vitro: HCC827, NCIH2135 and HUVEC cells	↓Tumoral cell viability	[208]
		In vivo: lung cancer animal model/intravenously	↓Body weight loss ↓Tumor volume ↑Tumor growth inhibition	
resveratrol	Galactosylated NP (NP combined with a ligand (galactose) for improved route of intestinal transport by the way of SGLT1)	In vitro: Raw 2647 cells	↓TNF-α, IL-6 and NO	[209]

↑: increase; ↓: decrease; CMC—carboxymethylcellulose; CXCL 1 and 2—chemokine ligand -1 and 2; DSS—dextran sulfate sodium; EGF—epidermal growth factor; HA—hyaluronic acid; IL—interleukin; Lcn.2—lipocalin-2;LPO—lipid peroxidation; MIA—mono-iodoacetate; MPO—myeloperoxidase; NO—nitric oxide; OA—osteoarthritis; OVA—ovalbumin; ROS—reactive oxidative species; SOD—superoxide dismutase; TNBS—trinitrobenzene sulfonic acid; TNF-α—tumoral necrosis factor-α; UC—ulcerative colitis; SGLT1—sodium glucose–linked transporter 1.

In animal models of ulcerative colitis (UC), codelivery of conventional drugs related to UC therapy, together with alternative therapeutic molecules or their combinations have been reported. Hyaluronic acid-functionalized polymeric NPs, to direct the specific drug (siCD98) and curcumin have shown anti-inflammatory effects in colonic epithelial cells and macrophages, protecting the mucosal layer and offering a structurally simple platform to be orally administered [203]. In that sense, a pH-sensitive NPs of curcumin-celecoxib combination reduces the overall toxicity and total dose of celecoxib, providing enhanced efficacy for mitigating UC by synergistic action of these two agents [204]. This novel form of carriers could represent a new strategy to deliver drugs directly to target cells in UC therapy.

As it is possible to observe, the use of NPs has been effective to improve the curcumin low systemic bioavailability. In a recent study, the protective effects of curcumin-loaded PLGA NPs, against mono-iodoacetate-induced osteoarthritis in rats, have been reported. The results reveal that curcumin could reverse hypocellularity and structural changes of articular cartilage in animal models of osteoarthritis. However, the increase in cellularity and matrix is more pronounced when it is encapsulated in PLGA [205].

Other nutraceuticals that offer health benefits have been nanoencapsulated to increase delivery, mobility, cellular uptake, bioaccessibility and stability. Carotenoids, widely distributed in fruit and vegetables, induce health beneficial properties mainly through their antioxidant activity, although their bioavailability is often compromised due to incomplete release from the food matrix, poor solubility and degradation during digestion [210]. Yi et al. [211] confirmed that whey protein isolate (WPI) NPs are good carriers for delivering beta-carotene, by means of the homogenization-evaporation method. This is due to the low release profile in gastric fluids and high release profile in intestinal fluids. Additionally, it has been observed that cellular antioxidant activity of beta-carotene improves with WPI-nanoencapsulation in CaCo-2 cells. Functional characteristics (such as antioxidative, antimutagenic, anticarcinogenic, antimicrobial properties) of green tea polyphenols are limited by their sensitivity to factors like temperature, light, pH, oxygen, etc. CH's NPs, a polysaccharide derived from chitin, can improve the bioaccessibility of tea derived phenols, by opening tight junctions and/or directly being absorbed by epithelial cells via endocytosis [212].

Most studies have failed to show a good bioavailability of many nutraceuticals, but the use of NPs may represent an alternative method to improve the beneficial effects, thus becoming a natural alternative treatment for several diseases.

5.2. *Toxicity*

Most commonly used nutraceuticals are compounds derived from herb food, plants, fruit and vegetables. Widely consumed nutraceuticals include flavonoid, flavonols and polyphenols, such as resveratrol, catechins and quercetin. A small number of these products do have a toxic potential, associated with hepatotoxicity, genotoxicity and mutagenicity [213,214]. In addition, the safety of some nutraceuticals can be compromised via contamination with toxic plants, metals, mycotoxins, pesticides, fertilizers or drug-supplement interactions [215]. Chemical structures of polyphenols could alleviate cytotoxicity induced by NPs through the inhibition of oxidative stress, hydrodynamic size, zeta potential and solubility caused by some NPs, such as the ones derived from zinc oxide (ZnO) [216]. The use of silver NPs (AgNPs) in several dietary supplements, utilized due to their strong antimicrobial properties, may leak out into the food and be consumed, creating severe health risks when reaching the small intestinal epithelium with their surface characteristics altered or dissolved into silver ions, which could alter their subsequent absorption and toxicity [217]. On the contrary, biopolymers, which are used for NP delivery systems, have well-documented biodegradable, biocompatible, mucoadhesive properties, and they do not decrease cellular viability in different cell lines when loaded with bioactive compounds [218].

The safety and beneficial properties on human health of nutraceuticals is well known (Figure 6). The toxicity and bioaccessibility and/or bioavailability could improve with the use of nanoparticle technology. Some NPs could actually cause potential cytotoxicity, the correct choice of nutraceutical-loaded NPs is important to deliver nutraceuticals and represent an alternative or complement to conventional medicine.

6. Future Challenges in DDS

The application of nanomedicine represents a huge breakthrough in the above-mentioned fields and assures an encouraging advance in the next decade. Treatments will become more efficient and safer due to the enormous variety of NP design and functionalization. The lists of potential applications progress to the point where the nanocarrier can be customized to best adjust to a certain active ingredient, a specific environment and then provide fitting drug location at the site of action, in a controlled manner. However, it is relevant to mention that NP-based treatments are not perfect and have challenges to conquer. First, the number of polymeric materials currently available for their utilization as DDS is still limited although the R&D has been moved in the last decade, exceeding expectations, from the micro- to the nanosize scale. The ideal adjustment to the delivery conditions, such as transportation to the site of action, specific targeting or adequate delivery profile, among others, for each type of disease, requires the development of new polymers that can fit these requisites. Although selective targeting supposed a great improvement in comparison to non-encapsulated drugs, it is a very complex mechanism and represents a challenge itself. Overexpression of a specific surface protein is not enough to assure selective targeting as they are also normally expressed in normal cells. This point is more critical in cancer treatments, where administered drugs usually possess higher toxicity that could lead to numerous undesirable secondary effects compared to drugs used in other diseases treatments. Most the assays have been developed in small animal models showing promising results, but the translation from animal results into clinical success has been limited. More clinical research and data are needed to fully comprehend the mechanism of these nanocarriers. In addition, limitations include the uncertain future of pharmaceutical companies which face high expenses concerning clinical trials and decreasing success rates in the flow of novel entities in the R&D pipeline. Examples of polymeric NPs that do not fulfil all the regulatory requirements for clinical evaluations and which had a harmful economic impact for their pharmaceutical companies are Livatag, PACA nanoparticulate formulation containing doxorubicin and BIND-014, PLGA polymer conjugated to docetaxel [219]. These formulations were potentially useful for the treatment of hepatocellular carcinoma and prostate cancer, respectively [219]. While BIND-014 began phase II of the clinical trial in 2018, the phase III studies of Livatag have not meet its primary endpoint of improving survival over, although it's action mechanism was demonstrated through DNA damage/synthesis inhibition and a decrease efflux pump by *P*-gp, due, at least in part, by an ion-pair association of doxorubicin with soluble degradation products of PACA which, conversely to free doxorubicin, are not a substrate for *P*-gp [220]. However, in both cases, no improvement was found when evaluated [221–223]. Perhaps focusing on more specific diseases, also considering aging population, novel formulations or indications for previous blockbusters drugs, including polymeric NPs, could be a good recommendation to maintain a profitable economic growth rate. Achieving reasonable success for oral bioavailability of poorly absorbed lipophilic and hydrophobic drugs, to maintain adequate and effective plasma levels over prolonged periods of time, still remains an important challenge. In addition, the fact that drugs used for severe illness are usually administered only through the parenteral route and the inaccessibility of most pharmacological targets are major constraints that are increasing interest in developing more efficient nanodelivery systems. Conceiving new methods for the manufacture of NPs at reasonable costs is an important part of this challenge because there are only a small number of them that fulfill the appropriated requirements to reach the target and to subsequently deliver the drug in a suitable manner. It is also mandatory for these polymeric NPs to be biodegradable or to possess a high capacity to be eliminated outside the body avoiding accumulation, being nontoxic and non-immunogenic. It is remarkable to point out the role that copolymers could play in tuning or modulating the interactions with mucosa or blood proteins in order to control their in vivo fate or to stabilize NPs without the need of surfactants entities. It would be also interesting for the near future research in this field to include stimuli responding polymers which can confer triggered release properties. From a manufacturing point of view, nanospheres and nanocapsules could be easily obtained by applying the existing methods, but new structures like polymersomes, are still waiting for better synthesis to join the family of nanoparticulate DDS. The need for developing NPs with many capabilities (targeting, image contrast enhancement), named as multifunctional NPs, means more synthetical steps, more regulatory hurdles and higher expenses. Conquering these objectives may seem very difficult, but there is hope of reaching a better scenario.

Over the last few years, there has been a global transformation in the field of nanomedicine, which has led to a multidisciplinary and collaborative approach with promising results and success. The future path of collaborations between theoretical and experimental scientists as well as the pharmaceutical industry, physicians and the regulatory agencies, will be crucial and will allow us to implement the laboratory results into the clinic and therefore, initiate the next generation of clinical therapies, trying to minimize the devastating consequences of terrible diseases such as pandemic covid-19.

In conclusion, many drawbacks or limitations still need to be resolved through numerous efforts and concentrated interdisciplinary scientific collaboration in order to reach the desired goals.

7. Conclusions

The toxicity associated with certain drugs and classical formulations or the complexity of treatment of some diseases, have driven the development of new alternatives as DDS. Among these, polymeric NPs are gaining high attention due to the biocompatibility, biodegradability and versatility they can offer, opening a wide range of materials that could possess the required characteristics for a specific application. For example, the use of hyaluronic acid in the NP outer surface increases adhesion to mucosal tissue and hence active ingredient liberation time, which is beneficial for drug delivery to eyes. Different techniques for cancer diagnosis are used with some disadvantages, such as the difficulty for early stage detection. The optimization of these techniques is possible due to different types of contrast agents, being NPs (e.g., gadolinium-based materials or AuNP) a promising agent in medical applications by the excellent biocompatibility, good water solubility and low toxicity. NP protection with PEG increases magnetic nanomaterials stability and avoids recognition by macrophages, which increases circulation time, which is indeed a requirement for diagnosis. At the same time, given that ABC transporters mediated MDR is the main obstacle for effective cancer therapy, the use of PEG as coating material for polymeric NPs has recently been described as an effective tool for inhibiting ABC transporters. The simultaneous use of one single NP for both cancer detection and drug delivery makes NPs a potential theranostic. Regarding NP pathways for drug delivery, passive diffusion, active targeting as well as stimuli responsive systems have been described. In this respect, the functionalization of NPs with the precise antibody, improves recognition of the specific site of action to achieve therapeutic effect, which drastically reduces secondary effects of drugs for oncologic treatments. In addition, the inclusion of highly unstable compounds used as nutraceuticals inside PVs, prevents them from being exposed to environments that could affect their integrity, implying an improvement in their absorption by the gastrointestinal system and hence, an increment of their bioavailability. This would suggest a new approach in nanomedicine for the use of nutraceuticals as an alternative or complementary treatment for different pathologies. Although important progress has been made in the fields of ocular drug delivery, cancer diagnosis and treatment and nutraceutical delivery, areas of medicine with an associated high level of morbidity, a notable reduction in the patient's quality of life or even an important mortality, in most cases, the translation from animal tests to real clinical success has been limited. The efforts applied in the development of new polymeric materials that may encompass the specific requirements for a certain delivery system, the better knowledge the scientists have about disease mechanisms and the collaborative research work carried out among all scientific areas, will boost the current state of the use of NPs in the medical field, which will be translated into more efficient and safer treatments.

Author Contributions: B.B. carried out the ocular drug delivery section and prepared the original draft; M.M. worked on the introduction; G.M. contributed with the diagnosis section; M.P.-A. conducted the DDS for cancer treatment section; T.O. and F.A.-A. prepared the nutraceuticals section and A.A. carried out the conceptualization and supervision of the work. All authors have read and agreed to the published version of the manuscript.

Funding: This research was supported by the Spanish Ministerio de Ciencia e Innovación (PID2019-109371GB-I00).

Acknowledgments: Authors would like to thank Rafael Prado-Gotor for the trust placed in us and Ana Pérez-Aranda for proofreading the ocular drug delivery section of the manuscript.

Conflicts of Interest: The authors declare no conflict of interest.

References

1. Bruschi, M.L. *Strategies to Modify the Drug Release from Pharmaceutical Systems*; Woodhead Publishing: Cambridge, UK, 2015; Capter 6; pp. 87–194. [CrossRef]
2. Jain, K.K. An Overview of Drug Delivery Systems. *Methods Mol. Biol.* **2020**, *2059*, 1–54. [CrossRef]
3. Macedo, A.S.; Castro, P.M.; Roque, L.; Thomé, N.G.; Reis, C.P.; Pintado, M.E.; Fonte, P. Novel and revisited approaches in nanoparticle systems for buccal drug delivery. *J. Control. Release* **2020**, *320*, 125–141. [CrossRef] [PubMed]

4. Jain, K.K. Role of Nanobiotechnology in Drug Delivery. *Methods Mol. Biol.* **2020**, *2059*, 55–73. [CrossRef] [PubMed]
5. Chen, Z.Y.; Wang, Y.X.; Lin, Y.; Zhang, J.S.; Yang, F.; Zhou, Q.L.; Liao, Y.Y. Advance of molecular imaging technology and targeted imaging agent in imaging and therapy. *Biomed Res. Int.* **2014**, *2014*, 1–12. [CrossRef] [PubMed]
6. Essa, D.; Kondiah, P.P.D.; Choonara, Y.E.; Pillay, V. The Design of Poly (lactide-co-glycolide) Nanocarriers for Medical Applications. *Front. Bioeng. Biotechnol.* **2020**, *8*, 48. [CrossRef]
7. Liu, R.; Zhao, J.; Han, Q.; Hu, X.; Wang, D.; Zhang, X.; Yang, P. One-Step Assembly of a Biomimetic Biopolymer Coating for Particle Surface Engineering. *Adv. Mater.* **2018**, *30*, e1802851. [CrossRef]
8. Pla, D.; Gómez, M. Metal and Metal Oxide Nanoparticles: A Lever for C–H Functionalization. *ACS Catal.* **2016**, *6*, 3537–3552. [CrossRef]
9. Begines, B.; de-Paz, M.-V.; Alcudia, A.; Galbis, J.A. Synthesis of reduction sensitive comb-like polyurethanes using click chemistry. *J. Polym. Sci. Part A Polym. Chem.* **2016**, *54*, 3888–3900. [CrossRef]
10. He, Y.F.; Zhang, F.; Saleh, E.; Vaithilingam, J.; Aboulkhair, N.; Begines, B.; Tuck, C.J.; Hague, R.J.M.; Ashcroft, I.A.; Wildman, R.D. A Tripropylene Glycol Diacrylate-based Polymeric Support Ink for Material Jetting. *Addit. Manuf.* **2017**, *16*, 153–161. [CrossRef]
11. Begines, B.; Zamora, F.; Violante de Paz, M.; Roffe, I.; Mancera, M.; Galbis, J.A. Synthesis and Characterization of New Carbohydrate-based Polyureas. *J. Renew. Mater.* **2013**, *1*, 212–221. [CrossRef]
12. Begines, B.; Alcudia, A.; Aguilera-Velazquez, R.; Martinez, G.; He, Y.; Wildman, R.; Sayagues, M.J.; Jimenez-Ruiz, A.; Prado-Gotor, R. Design of highly stabilized nanocomposite inks based on biodegradable polymer-matrix and gold nanoparticles for Inkjet Printing. *Sci. Rep.* **2019**, *9*, 16097. [CrossRef] [PubMed]
13. Nakama, Y. Surfactants. In *Cosmetic Science and Technology. Theoretical Principles and Applications*; Elsevier: Amsterdam, The Netherlands, 2017; pp. 231–244.
14. Zhao, T.; Elzatahry, A.; Li, X.; Zhao, D. Single-micelle-directed synthesis of mesoporous materials. *Nat. Rev. Mater.* **2019**, *4*, 775–791. [CrossRef]
15. Belletti, D.; Grabrucker, A.M.; Pederzoli, F.; Menrath, I.; Cappello, V.; Vandelli, M.A.; Forni, F.; Tosi, G.; Ruozi, B. Exploiting The Versatility of Cholesterol in Nanoparticles Formulation. *Int. J. Pharm.* **2016**, *511*, 331–340. [CrossRef] [PubMed]
16. Bonaccorso, A.; Musumeci, T.; Carbone, C.; Vicari, L.; Lauro, M.R.; Puglisi, G. Revisiting the role of sucrose in PLGA-PEG nanocarrier for potential intranasal delivery. *Pharm. Dev. Technol.* **2018**, *23*, 265–274. [CrossRef] [PubMed]
17. Joseph, E.; Saha, R.N. Investigations on pharmacokinetics and biodistribution of polymeric and solid lipid nanoparticulate systems of atypical antipsychotic drug: Effect of material used and surface modification. *Drug Dev. Ind. Pharm.* **2017**, *43*, 678–686. [CrossRef] [PubMed]
18. Gao, W.; Chen, Y.; Thompson, D.H.; Park, K.; Li, T. Impact of surfactant treatment of paclitaxel nanocrystals on biodistribution and tumor accumulation in tumor-bearing mice. *J. Control. Release* **2016**, *237*, 168–176. [CrossRef] [PubMed]
19. Chen, Z.; Shi, T.; Zhang, L.; Zhu, P.; Deng, M.; Huang, C.; Hu, T.; Jiang, L.; Li, J. Mammalian drug efflux transporters of the ATP binding cassette (ABC) family in multidrug resistance: A review of the past decade. *Cancer Lett.* **2016**, *370*, 153–164. [CrossRef]
20. Yin, J.; Deng, X.; Zhang, J.; Lin, J. Current Understanding of Interactions between Nanoparticles and ABC Transporters in Cancer Cells. *Curr. Med. Chem.* **2018**, *25*, 5930–5944. [CrossRef]
21. World Health Organization. Blindness and Vision Impairment. Available online: https://www.who.int/en/news-room/fact-sheets/detail/blindness-and-visual-impairment (accessed on 20 March 2020).
22. Souto, E.B.; Dias-Ferreira, J.; López-Machado, A.; Ettcheto, M.; Sánchez-López, E. Advanced Formulation Approaches for Ocular Drug Delivery: State-Of-The-Art and Recent Patents. *Pharmaceutics* **2019**, *11*, 460. [CrossRef]
23. Neumann, R.; Barequet, D. The gap between the need for novel retinal drug delivery methods, technologies in R&D phase, and approved ocular drug delivery technologies. *Drug Discov. Today* **2019**, *24*, 1433–1435. [CrossRef]
24. Meng, T.; Kulkarni, V.; Simmers, R.; Brar, V.; Xu, Q. Therapeutic implications of nanomedicine for ocular drug delivery. *Drug Discov. Today* **2019**, *24*, 1524–1538. [CrossRef]

25. Gote, V.; Sikder, S.; Sicotte, J.; Pal, D. Ocular Drug Delivery: Present Innovations and Future Challenges. *J. Pharmacol. Exp. Ther.* **2019**, *370*, 602–624. [CrossRef]
26. Juliana, F.R.; Kesse, S.; Boakye-Yiadom, K.O.; Veroniaina, H.; Wang, H.; Sun, M. Promising Approach in the Treatment of Glaucoma Using Nanotechnology and Nanomedicine-Based Systems. *Molecules* **2019**, *24*, 3805. [CrossRef] [PubMed]
27. Blazaki, S.; Pachis, K.; Tzatzarakis, M.; Tsilimbaris, M.; Antimisiaris, S.G. Novel Liposome Aggregate Platform (LAP) system for sustained retention of drugs in the posterior ocular segment following intravitreal injection. *Int. J. Pharm.* **2020**, *576*, 118987. [CrossRef]
28. Dukovski, B.J.; Juretić, M.; Bračko, D.; Randjelović, D.; Savić, S.; Crespo Moral, M.; Diebold, Y.; Filipović-Grčić, J.; Pepić, I.; Lovrić, J. Functional ibuprofen-loaded cationic nanoemulsion: Development and optimization for dry eye disease treatment. *Int. J. Pharm.* **2020**, *576*, 118979. [CrossRef]
29. Kim, H.-J.; Zhang, K.; Moore, L.; Ho, D. Diamond Nanogel-Embedded Contact Lenses Mediate Lysozyme-Dependent Therapeutic Release. *ACS Nano* **2014**, *8*, 2998–3005. [CrossRef]
30. Romero, G.B.; Keck, C.M.; Müller, R.H.; Bou-Chacra, N.A. Development of cationic nanocrystals for ocular delivery. *Eur. J. Pharm. Biopharm.* **2016**, *107*, 215–222. [CrossRef]
31. Chi, H.; Gu, Y.; Xu, T.; Cao, F. Multifunctional organic–inorganic hybrid nanoparticles and nanosheets based on chitosan derivative and layered double hydroxide: Cellular uptake mechanism and application for topical ocular drug delivery. *Int. J. Nanomed.* **2017**, *12*, 1607–1620. [CrossRef]
32. Park, K.; Skidmore, S.; Hadar, J.; Garner, J.; Park, H.; Otte, A.; Soh, B.K.; Yoon, G.; Yu, D.; Yun, Y.; et al. Injectable, long-acting PLGA formulations: Analyzing PLGA and understanding microparticle formation. *J. Control. Release* **2019**, *304*, 125–134. [CrossRef]
33. Varshochian, R.; Riazi-Esfahani, M.; Jeddi-Tehrani, M.; Mahmoudi, A.-R.; Aghazadeh, S.; Mahbod, M.; Movassat, M.; Atyabi, F.; Sabzevari, A.; Dinarvand, R. Albuminated PLGA nanoparticles containing bevacizumab intended for ocular neovascularization treatment. *J. Biomed. Mater. Res. A* **2015**, *103A*, 3148–3156. [CrossRef]
34. Ryu, W.M.; Kim, S.-N.; Min, C.H.; Choy, Y.B. Dry Tablet Formulation of PLGA Nanoparticles with a Preocular Applicator for Topical Drug Delivery to the Eye. *Pharmaceutics* **2019**, *11*, 651. [CrossRef]
35. Qiu, F.; Meng, T.; Chen, Q.; Zhou, K.; Shao, Y.; Matlock, G.; Ma, X.; Wu, W.; Du, Y.; Wang, X.; et al. Fenofibrate-Loaded Biodegradable Nanoparticles for the Treatment of Experimental Diabetic Retinopathy and Neovascular Age-RelatedMacular Degeneration. *Molecualr Pharm.* **2019**, *16*, 1958–1970. [CrossRef]
36. Liu, J.; Zhang, X.; Li, G.; Xu, F.; Li, S.; Teng, L.; Li, Y.; Sun, F. Anti-Angiogenic Activity Of Bevacizumab-Bearing Dexamethasone-Loaded PLGA Nanoparticles ForPotential Intravitreal Applications. *Int. J. Nanomed.* **2019**, *14*, 8819–8834. [CrossRef]
37. Salama, H.A.; Ghorab, M.; Mahmoud, A.A.; Abdel Hady, M. PLGA Nanoparticles as Subconjunctival Injection for Management of Glaucoma. *AAPS PharmSciTech* **2017**, *18*, 2517–2528. [CrossRef]
38. Pan, Q.; Xu, Q.; Boylan, N.J.; Lamb, N.W.; Emmert, D.G.; Yang, J.-C.; Tang, L.; Heflin, T.; Alwadani, S.; Eberhart, C.G.; et al. Corticosteroid-loaded biodegradable nanoparticles for prevention of corneal allograft rejection in rats. *J. Control. Release* **2015**, *201*, 32–40. [CrossRef]
39. Li, F.; Hurley, B.; Liu, Y.; Leonard, B.; Griffith, M. Controlled Release of Bevacizumab through Nanospheres for Extended Treatment of Age-Related Macular Degeneration. *Open Ophthalmol. J.* **2012**, *6*, 54–58. [CrossRef]
40. Imam, S.S.; Bukhari SN, A.; Ahmad, J.; Ali, A. Formulation and optimization of levofloxacin loaded chitosannanoparticle for ocular delivery: In-vitro characterization, oculartolerance and antibacterial activity. *Int. J. Biol. Macromol.* **2018**, *108*, 650–659. [CrossRef]
41. Badiee, P.; Varshochian, R.; Rafiee-Tehrani, M.; Abedin Dorkoosh, F.; Khoshayand, M.R.; Dinarvand, R. Ocular implant containing bevacizumab-loaded chitosan nanoparticles intended for choroidal neovascularization treatment. *J. Biomed. Mater. Res. A* **2018**, *106*, 2261–2271. [CrossRef]
42. Silva, M.M.; Calado, R.; Marto, J.; Bettencourt, A.; Almeida, A.J.; Gonçalves, L.M.D. Chitosan Nanoparticles as a Mucoadhesive Drug Delivery System for Ocular Administration. *Mar. Drugs* **2017**, *15*, 370. [CrossRef]
43. Hanafy, A.F.; Abdalla, A.M.; Guda, T.K.; Gabr, K.E.; Royall, P.G.; Alqurshi, A. Ocular anti-inflammatory activity of prednisolone acetate loaded chitosan-deoxycholate self-assembled nanoparticles. *Int. J. Nanomed.* **2019**, *14*, 3679–3689. [CrossRef]
44. Xu, X.; Sun, L.; Zhou, L.; Cheng, Y.; Cao, F. Functional chitosan oligosaccharide nanomicelles for topical ocular drugdelivery of dexamethasone. *Carbohydr. Polym.* **2020**, *227*. [CrossRef]

45. Shi, S.; Zhang, Z.; Luo, Z.; Yu, J.; Liang, R.; Li, X.; Chen, H. Chitosan grafted methoxy poly(ethylene glycol)-poly(e-caprolactone) nanosuspension for ocular delivery of hydrophobic diclofenac. *Sci. Rep.* **2015**, *5*, 11337. [CrossRef]
46. Yu, Y.; Chen, D.; Li, Y.; Yang, W.; Tu, J.; Shen, Y. Improving the topical ocular pharmacokinetics of lyophilized cyclosporine A-loaded micelles: Formulation, in vitro and in vivo studies. *Drug Deliv.* **2019**, *25*, 888–899. [CrossRef]
47. Tang, Z.; Yin, L.; Zhang, Y.; Yu, W.; Wang, Q.; Zhan, Z. Preparation and study of two kinds of ophthalmic nano-preparations of everolimus. *Drug Deliv.* **2019**, *26*, 1235–1242. [CrossRef]
48. Huang, J.; Yu, X.; Zhou, Y.; Zhang, R.; Song, Q.; Wang, Q.; Li, X. Directing the nanoparticle formation by the combination with small molecular assembly and polymeric assembly for topical suppression of ocular inflammation. *Int. J. Pharm.* **2018**, *551*, 223–231. [CrossRef] [PubMed]
49. Wen, Y.; Ban, J.; Mo, Z.; Zhang, Y.; An, P.; Liu, L.; Xie, Q.; Du, Y.; Xie, B.; Zhan, X.; et al. A potential nanoparticle-loaded in situ gel for enhanced and sustained ophthalmic delivery of dexamethasone. *Nanotechnology* **2018**, *29*, 425101. [CrossRef]
50. Hirani, A.; Grover, A.; Lee, Y.W.; Pathak, Y.; Sutariya, V. Triamcinolone acetonide nanoparticles incorporated in thermoreversible gels for agerelated macular degeneration. *Pharm. Dev. Technol.* **2016**, *21*, 61–67. [CrossRef]
51. Yandrapu, S.K.; Upadhyay, A.K.; Petrash, J.M.; Kompella, U.B. Nanoparticles in Porous Microparticles Prepared by Supercritical Infusion and Pressure Quench Technology for Sustained Delivery of Bevacizumab. *Mol. Pharm.* **2012**, *10*, 4676–4686. [CrossRef]
52. Yang, X.; Wang, L.; Li, L.; Han, M.; Tang, S.; Wang, T.; Han, J.; He, X.; He, X.; Wanga, A.; et al. A novel dendrimer-based complex co-modified with cyclic RGD hexapeptide and penetratin for noninvasive targeting and penetration of the ocular posterior segment. *Drug Deliv.* **2019**, *26*, 989–1001. [CrossRef]
53. Lancina, M.G.; Wang, J.; Williamson, G.S.; Yang, H. DenTimol as A Dendrimeric Timolol Analogue for Glaucoma Therapy: Synthesis and Preliminary Efficacy and Safety Assessment. *Mol. Pharm.* **2018**, *15*, 2883–2889. [CrossRef]
54. Tai, L.; Liu, C.; Jiang, K.; Chen, X.; Feng, L.; Pan, W.; Wei, G.; Lu, W. A novel penetratin-modified complex for noninvasive intraocular delivery of antisense oligonucleotides. *Int. J. Pharm.* **2017**, *529*, 347–356. [CrossRef] [PubMed]
55. Kambhampati, S.P.; Kannan, R.M. Dendrimer Nanoparticles for Ocular Drug Delivery. *J. Ocul. Pharmacol. Ther.* **2013**, *29*, 151–165. [CrossRef] [PubMed]
56. Wu, X.; Yu, G.; Luo, C.; Maeda, A.; Zhang, N.; Sun, D.; Zhou, Z.; Puntel, A.; Palczewski, K.; Lu, Z.-R. Synthesis and evaluation of a nanoglobular dendrimer 5- aminosalicylic acid conjugate with a hydrolyzable chiff base spacer for treating retinal degeneration. *ACS Nano* **2014**, *8*, 153–161. [CrossRef]
57. Spataro, G.; Malecaze, F.; Turrin, C.-O.; Soler, V.; Duhayon, C.; Elena, P.-P.; Majoral, J.-P.; Caminade, A.-M. Designing dendrimers for ocular drug delivery. *Eur. J. Med. Chem.* **2010**, *45*, 326–334. [CrossRef] [PubMed]
58. Tian, B.; Hua, S.; Liu, J. Cyclodextrin-based delivery systems for chemotherapeutic anticancer drugs: A review. *Carbohydr. Polym.* **2020**, *232*, 115805. [CrossRef]
59. Wankar, J.; Kotla, N.G.; Gera, S.; Rasala, S.; Pandit, A.; Rochev, Y.A. Implications for Responsive Drug Delivery and Biomedical Engineering. *Adv. Funct. Mater.* **2020**, 1909049. [CrossRef]
60. Rodriguez-Aller, M.; Guinchard, S.; Guillarme, D.; Pupier, M.; Jeannerat, D.; Rivara-Minten, E.; Veuthey, J.L.; Gurny, R. New prostaglandin analog formulation for glaucoma treatment containing cyclodextrins for improved stability, solubility and ocular tolerance. *Eur. J. Pharm. Biopharm.* **2015**, *95*, 203–214. [CrossRef]
61. Jansook, P.; Kulsirachote, P.; Loftsson, T. Cyclodextrin solubilization of celecoxib: Solid and solution state characterization. *J. Incl. Phenom. Macrocycl. Chem.* **2018**, *90*, 75–88. [CrossRef]
62. Jansook, P.; Kulsirachote, P.; Asasutjarit, R.; Loftsson, T. Development of celecoxib eye drop solution and microsuspension: A comparative investigation of binary and ternary cyclodextrin complexes. *Carbohydr. Polym.* **2019**, *225*, 115209. [CrossRef]
63. Lorenzo-Veiga, B.; Sigurdsson, H.H.; Loftsson, T.; Alvarez-Lorenzo, C. Cyclodextrin-Amphiphilic Copolymer Supramolecular Assemblies for the Ocular Delivery of Natamycin. *Nanomaterials* **2019**, *9*, 745. [CrossRef]
64. Farace, C.; Sánchez-Moreno, P.; Orecchioni, M.; Manetti, R.; Sgarrella, F.; Asara, Y.; Peula-García, J.M.; Marchal, J.A.; Madeddu, R.; Delogu, L.G. Immune cell impact of three differently coated lipid nanocapsules: Pluronic, chitosan and polyethylene glycol. *Sci. Rep.* **2016**, *6*, 18423. [CrossRef]

65. Zorzi, G.K.; Schuh, R.S.; Maschio, V.J.; Brazil, N.T.; Rott, M.B.; Teixeira, H.F. Box Behnken design of siRNA-loaded liposomes for the treatment of amurine model of ocular keratitis caused byAcanthamoeba. *Colloids Surf. B Biointerfaces* **2019**, *173*, 725–732. [CrossRef]
66. Patil, A.; Lakhani, P.; Taskar, P.; Wu, K.-W.; Sweeney, C.; Avula, B.; Wang, Y.-H.; Khan, I.A.; Majumdar, S. Formulation Development, Optimization, and In vitro—In vivo Characterization of Natamycin Loaded PEGylated Nano-lipid Carriers for Ocular Applications. *J. Pharm. Sci.* **2018**, *107*, 2160–2171. [CrossRef]
67. Fathi Karkan, S.; Mohammadhosseini, M.; Panahi, Y.; Milani, M.; Zarghami, N.; Akbarzadeh, A.; Abasi, E.; Hosseini, A.; Davaran, S. Magnetic nanoparticles in cancer diagnosis and treatment: A review. *Artif. Cells Nanomed. Biotechnol.* **2017**, *45*, 1–5. [CrossRef]
68. Zhou, X.Y.; Tay, Z.W.; Chandrasekharan, P.; Yu, E.Y.; Hensley, D.W.; Orendorff, R.; Jeffris, K.E.; Mai, D.; Zheng, B.; Goodwill, P.W.; et al. Magnetic particle imaging for radiation-free, sensitive and high-contrast vascular imaging and cell tracking. *Curr. Opin. Chem. Biol.* **2018**, *45*, 131–138. [CrossRef]
69. Wang, Y.; Xia, Y. Optical, electrochemical and catalytic methods for in-vitro diagnosis using carbonaceous nanoparticles: A review. *Microchim. Acta* **2019**, *186*, 50. [CrossRef]
70. Thakor, A.S.; Gambhir, S.S. Nanooncology: The future of cancer diagnosis and therapy. *CA Cancer J. Clin.* **2013**, *63*, 395–418. [CrossRef]
71. Maham, M.M.; Doiron, A.L. Gold Nanoparticles as X-Ray, CT, and Multimodal Imaging Contrast Agents: Formulation, Targeting, and Methodology. *J. Nanomater.* **2018**, *2018*, 5837276.
72. Kim, D.; Park, S.; Lee, J.H.; Jeong, Y.Y.; Jon, S. Antibiofouling Polymer-Coated Gold Nanoparticles as a Contrast Agent for in Vivo X-ray Computed Tomography Imaging. *J. Am. Chem. Soc.* **2007**, *129*, 7661–7665. [CrossRef]
73. Rabin, O.; Manuel Perez, J.; Grimm, J.; Wojtkiewicz, G.; Weissleder, R. An X-ray computed tomography imaging agent based on long-circulating bismuth sulphide nanoparticles. *Nat. Mater.* **2006**, *5*, 118–122. [CrossRef]
74. Al Zaki, A.; Joh, D.; Cheng, Z.; De Barros, A.L.B.; Kao, G.; Dorsey, J.; Tsourkas, A. Gold-Loaded Polymeric Micelles for Computed Tomography-Guided Radiation Therapy Treatment and Radiosensitization. *ACS Nano* **2014**, *8*, 104–112. [CrossRef] [PubMed]
75. Lin, W.J.; Zhang, X.F.; Qian, L.; Yao, N.; Pan, Y.; Zhang, L.J. Doxorubicin-Loaded Unimolecular Micelle-Stabilized Gold Nanoparticles as a Theranostic Nanoplatform for Tumor-Targeted Chemotherapy and Computed Tomography Imaging. *Biomacromolecules* **2017**, *18*, 3869–3880. [CrossRef] [PubMed]
76. Park, J.H.; Park, J.; Kim, S.; Kim, S.-H.; Lee, T.G.; Lee, J.Y.; Wi, J.-S. Characterization and application of porous gold nanoparticles as 2-photon luminescence imaging agents: 20-fold brighter than gold nanorods. *J. Biophotonics* **2018**, *11*, e201700174. [CrossRef] [PubMed]
77. Fernández, T.D.; Pearson, J.R.; Leal, M.P.; Torres, M.J.; Blanca, M.; Mayorga, C.; Le Guével, X. Intracellular accumulation and immunological properties of fluorescent gold nanoclusters in human dendritic cells. *Biomaterials* **2015**, *43*, 1–12. [CrossRef] [PubMed]
78. Wen, T.; He, Y.; Liu, X.-L.; Lin, M.-L.; Cheng, Y.; Zhao, J.; Gong, Q.; Xia, K.; Tan, P.-H.; Lu, G. Spectral shape of one-photon luminescence from single gold nanorods. *AIP Adv.* **2017**, *7*, 125106. [CrossRef]
79. Wang, Y.; Strohm, E.M.; Sun, Y.; Wang, Z.; Zheng, Y.; Wang, Z.; Kolios, M.C. Biodegradable polymeric nanoparticles containing gold nanoparticles and Paclitaxel for cancer imaging and drug delivery using photoacoustic methods. *Biomed. Opt. Express* **2016**, *7*, 4125–4138. [CrossRef]
80. Wang, L.; Hu, S. Photoacoustic Tomography: In Vivo Imaging from Organelles to Organs. *Science* **2012**, *335*, 1458–1462. [CrossRef]
81. Weissleder, R. Molecular Imaging in Cancer. *Science* **2006**, *312*, 1168. [CrossRef]
82. Caravan, P. Strategies for increasing the sensitivity of gadolinium based MRI contrast agents. *Chem. Soc. Rev.* **2006**, *35*, 512–523. [CrossRef]
83. Chauhan, V.P.; Jain, R.K. Strategies for advancing cancer nanomedicine. *Nat. Mater.* **2013**, *12*, 958–962. [CrossRef]
84. Choi, H.S.; Liu, W.; Liu, F.; Nasr, K.; Misra, P.; Bawendi, M.G.; Frangioni, J.V. Design considerations for tumour-targeted nanoparticles. *Nat. Nanotechnol.* **2010**, *5*, 42–47. [CrossRef]
85. Ratzinger, G.; Agrawal, P.; Körner, W.; Lonkai, J.; Sanders, H.M.H.F.; Terreno, E.; Wirth, M.; Strijkers, G.J.; Nicolay, K.; Gabor, F. Surface modification of PLGA nanospheres with Gd-DTPA and Gd-DOTA for high-relaxivity MRI contrast agents. *Biomaterials* **2010**, *31*, 8716–8723. [CrossRef]

86. Caravan, P.; Zhang, Z. Targeted MRI Contrast Agents. *Chem. Contrast Agents Med Magn. Reson. Imaging* **2013**. [CrossRef]
87. Liu, Y.; Chen, Z.; Liu, C.; Yu, D.; Lu, Z.; Zhang, N. Gadolinium-loaded polymeric nanoparticles modified with Anti-VEGF as multifunctional MRI contrast agents for the diagnosis of liver cancer. *Biomaterials* **2011**, *32*, 5167–5176. [CrossRef]
88. Esser, L.; Truong, N.P.; Karagoz, B.; Moffat, B.A.; Boyer, C.; Quinn, J.F.; Whittaker, M.R.; Davis, T.P. Gadolinium-functionalized nanoparticles for application as magnetic resonance imaging contrast agents via polymerization-induced self-assembly. *Polym. Chem.* **2016**, *7*, 7325–7337. [CrossRef]
89. Kim, C.; Song, K.H.; Gao, F.; Wang, L.V. Sentinel lymph nodes and lymphatic vessels: Noninvasive dual-modality in vivo mapping by using indocyanine green in rats—Volumetric spectroscopic photoacoustic imaging and planar fluorescence imaging. *Radiology* **2010**, *255*, 442–450. [CrossRef]
90. Song, K.H.; Stein, E.W.; Margenthaler, J.A.; Wang, L.V. Noninvasive photoacoustic identification of sentinel lymph nodes containing methylene blue in vivo in a rat model. *J. Biomed. Opt.* **2008**, *13*, 054033. [CrossRef]
91. Li, Y.; Beija, M.; Laurent, S.; Vander Elst, L.; Muller, R.; Duong, H.; Lowe, A.; Davis, T.; Boyer, C. Macromolecular Ligands for Gadolinium MRI Contrast Agents. *Macromolecules* **2012**, *45*, 4196–4204. [CrossRef]
92. Zhang, G.; Zhang, R.; Wen, X.; Li, L.; Li, C. Micelles based on biodegradable poly (L-glutamic acid)-b-polylactide with paramagnetic Gd ions chelated to the shell layer as a potential nanoscale MRI-visible delivery system. *Biomacromolecules* **2008**, *9*, 36–42. [CrossRef]
93. Hu, J.; Liu, T.; Zhang, G.; Jin, F.; Liu, S. Synergistically Enhance Magnetic Resonance/Fluorescence Imaging Performance of Responsive Polymeric Nanoparticles under Mildly Acidic Biological Milieu. *Macromol. Rapid Commun.* **2013**, *34*, 749–758. [CrossRef]
94. Li, A.; Luehmann, H.P.; Sun, G.; Samarajeewa, S.; Zou, J.; Zhang, S.; Zhang, F.; Welch, M.J.; Liu, Y.; Wooley, K.L. Synthesis and In Vivo Pharmacokinetic Evaluation of Degradable Shell Cross-Linked Polymer Nanoparticles with Poly(carboxybetaine) versus Poly(ethylene glycol) Surface-Grafted Coatings. *ACS Nano* **2012**, *6*, 8970–8982. [CrossRef] [PubMed]
95. Shahbazi-Gahrouei, D.; Williams, M.; Rizvi, S.; Allen, B.J. In vivo studies of Gd-DTPA-monoclonal antibody and gd-porphyrins: Potential magnetic resonance imaging contrast agents for melanoma. *J. Magn. Reson. Imaging* **2001**, *14*, 169–174. [CrossRef]
96. Hu, X.; Lu, F.; Chen, L.; Tang, Y.; Hu, W.; Lu, X.; Ji, Y.; Yang, Z.; Zhang, W.; Yin, C.; et al. Perylene Diimide-Grafted Polymeric Nanoparticles Chelated with Gd3+ for Photoacoustic/T1-Weighted Magnetic Resonance Imaging-Guided Photothermal Therapy. *ACS Appl. Mater. Interfaces* **2017**, *9*, 30458–30469. [CrossRef] [PubMed]
97. Wu, B.; Lu, S.-T.; Yu, H.; Liao, R.-F.; Li, H.; Lucie Zafitatsimo, B.V.; Li, Y.-S.; Zhang, Y.; Zhu, X.-L.; Liu, H.-G.; et al. Gadolinium-chelate functionalized bismuth nanotheranostic agent for in vivo MRI/CT/PAI imaging-guided photothermal cancer therapy. *Biomaterials* **2018**, *159*, 37–47. [CrossRef] [PubMed]
98. Ruiz-Cabello, J.; Barnett, B.P.; Bottomley, P.A.; Bulte, J.W.M. Fluorine (19F) MRS and MRI in biomedicine. *NMR Biomed.* **2011**, *24*, 114–129. [CrossRef]
99. Cyrus, T.; Winter, P.M.; Caruthers, S.D.; Wickline, S.A.; Lanza, G.M. Magnetic resonance nanoparticles for cardiovascular molecular imaging and therapy. *Expert Rev. Cardiovasc. Ther.* **2005**, *3*, 705–715. [CrossRef] [PubMed]
100. Knight, J.C.; Edwards, P.G.; Paisey, S.J. Fluorinated contrast agents for magnetic resonance imaging; a review of recent developments. *RSC Adv.* **2011**, *1*, 1415–1425. [CrossRef]
101. Wek, K.S. Development of Polymeric Therapeutic Nanoparticles: Toward Targeted Delivery and Efficient ^{19}F MRI of Solid Tumors. Ph.D. Thesis, Case Western Reserve University, Cleveland, OH, USA, 2017.
102. Wallat, J.D.; Czapar, A.E.; Wang, C.; Wen, A.M.; Wek, K.S.; Yu, X.; Steinmetz, N.F.; Pokorski, J.K. Optical and Magnetic Resonance Imaging Using Fluorous Colloidal Nanoparticles. *Biomacromolecules* **2017**, *18*, 103–112. [CrossRef]
103. Pisani, E.; Tsapis, N.; Paris, J.; Nicolas, V.; Cattel, L.; Fattal, E. Polymeric Nano/Microcapsules of Liquid Perfluorocarbons for Ultrasonic Imaging: Physical Characterization. *Langmuir* **2006**, *22*, 4397–4402. [CrossRef]

104. Giraudeau, C.; Flament, J.; Marty, B.; Boumezbeur, F.; Mériaux, S.; Robic, C.; Port, M.; Tsapis, N.; Fattal, E.; Giacomini, E.; et al. A new paradigm for high-sensitivity 19F magnetic resonance imaging of perfluorooctylbromide. *Magn. Reson. Med.* **2010**, *63*, 1119–1124. [CrossRef]
105. Diou, O.; Fattal, E.; Delplace, V.; Mackiewicz, N.; Nicolas, J.; Mériaux, S.; Valette, J.; Robic, C.; Tsapis, N. RGD decoration of PEGylated polyester nanocapsules of perfluorooctyl bromide for tumor imaging: Influence of pre or post-functionalization on capsule morphology. *Eur. J. Pharm. Biopharm.* **2014**, *87*, 170–177. [CrossRef] [PubMed]
106. Liopo, A.; Su, R.; Oraevsky, A.A. Melanin nanoparticles as a novel contrast agent for optoacoustic tomography. *Photoacoustics* **2015**, *3*, 35–43. [CrossRef] [PubMed]
107. Belletti, D.; Riva, G.; Luppi, M.; Tosi, G.; Forni, F.; Vandelli, M.; Pederzoli, F. Anticancer drug-loaded quantum dots engineered polymeric nanoparticles: Diagnosis/therapy combined approach. *Eur. J. Pharm. Sci.* **2017**, *107*. [CrossRef]
108. Zhou, B.; Xiong, Z.; Wang, P.; Peng, C.; Shen, M.; Mignani, S.; Majoral, J.-P.; Shi, X. Targeted tumor dual mode CT/MR imaging using multifunctional polyethylenimine-entrapped gold nanoparticles loaded with gadolinium. *Drug Deliv.* **2018**, *25*, 178–186. [CrossRef]
109. McQuade, C.; Al Zaki, A.; Desai, Y.; Vido, M.; Sakhuja, T.; Cheng, Z.; Hickey, R.J.; Joh, D.; Park, S.-J.; Kao, G.; et al. A Multifunctional Nanoplatform for Imaging, Radiotherapy, and the Prediction of Therapeutic Response. *Small* **2015**, *11*, 834–843. [CrossRef]
110. Topete, A.; Alatorre-Meda, M.; Villar-Alvarez, E.M.; Carregal-Romero, S.; Barbosa, S.; Parak, W.J.; Taboada, P.; Mosquera, V. Polymeric-Gold Nanohybrids for Combined Imaging and Cancer Therapy. *Adv. Healthc. Mater.* **2014**, *3*, 1309–1325. [CrossRef]
111. Sun, J.; Zhou, S.; Hou, P.; Yang, Y.; Weng, J.; Li, X.; Li, M. Synthesis and characterization of biocompatible Fe_3O_4 nanoparticles. *J. Biomed. Mater. Res. Part A* **2007**, *80A*, 333–341. [CrossRef]
112. Qiao, R.; Yang, C.; Gao, M. Superparamagnetic iron oxide nanoparticles: From preparations to in vivo MRI applications. *J. Mater. Chem.* **2009**, *19*, 6274–6293. [CrossRef]
113. Fang, C.; Zhang, M. Multifunctional magnetic nanoparticles for medical imaging applications. *J. Mater. Chem.* **2009**, *19*, 6258–6266. [CrossRef]
114. Laurent, S.; Forge, D.; Port, M.; Roch, A.; Robic, C.; Vander Elst, L.; Muller, R.N. Magnetic Iron Oxide Nanoparticles: Synthesis, Stabilization, Vectorization, Physicochemical Characterizations, and Biological Applications. *Chem. Rev.* **2008**, *108*, 2064–2110. [CrossRef]
115. Zhao, C.; Song, X.; Jin, W.; Wu, F.; Zhang, Q.; Zhang, M.; Zhou, N.; Shen, J. Image-guided cancer therapy using aptamer-functionalized cross-linked magnetic-responsive Fe_3O_4@carbon nanoparticles. *Anal. Chim. Acta* **2019**, *1056*, 108–116. [CrossRef]
116. Luo, Y.; Yang, J.; Yan, Y.; Li, J.; Shen, M.; Zhang, G.; Mignani, S.; Shi, X. RGD-functionalized ultrasmall iron oxide nanoparticles for targeted T1-weighted MR imaging of gliomas. *Nanoscale* **2015**, *7*, 14538–14546. [CrossRef] [PubMed]
117. Pernia Leal, M.; Rivera-Fernández, S.; Franco, J.M.; Pozo, D.; de la Fuente, J.M.; García-Martín, M.L. Long-circulating PEGylated manganese ferrite nanoparticles for MRI-based molecular imaging. *Nanoscale* **2015**, *7*, 2050–2059. [CrossRef] [PubMed]
118. Calzoni, E.; Cesaretti, A.; Polchi, A.; Di Michele, A.; Tancini, B.; Emiliani, C. Biocompatible Polymer Nanoparticles for Drug Delivery Applications in Cancer and Neurodegenerative Disorder Therapies. *J. Funct. Biomater.* **2019**, *10*, 4. [CrossRef] [PubMed]
119. Tran, S.; DeGiovanni, P.; Piel, B.; Rai, P. Cancer nanomedicine: A review of recent success in drug delivery. *Clin. Trans. Med.* **2017**, *6*. [CrossRef]
120. Bregoli, L.; Movia, D.; Gavigan-Imedio, J.; Lysaght, J.; Reynolds, J.; Prina-Mello, A. Nanomedicine applied to translational oncology: A future perspective on cancer treatment. *Nanomedicine* **2016**, *12*, 81–103. [CrossRef] [PubMed]
121. Priyamvada, P.; Chandana, M.; Sahoo, S. Nanotechnology-based combinational drug delivery:an emerging approach for cancer therapy. *Drug Discos. Today* **2012**, *17*. [CrossRef]
122. Cho, K.; Wang, X.U.; Nie, S.; Shin, D.M. Therapeutic Nanoparticles for Drug Delivery in Cancer. *Clin. Cancer Res.* **2008**, *14*. [CrossRef]
123. Senapati, S.; Mahanta, A.K.; Kumar, S.; Maiti, P. Controlled drug delivery vehicles for cancer treatment and their performance. *Signal Transduct. Target. Ther.* **2018**, *3*, 7. [CrossRef]

124. Jiang, Y.; Wong, S.; Chen, F.; Chang, T.; Lu, H.; Stenzel, M.H. Influencing Selectivity to Cancer Cells with Mixed Nanoparticles Prepared from Albumin–Polymer Conjugates and Block Copolymers. *Bioconjugate Chem.* **2017**, *28*, 979–985. [CrossRef]
125. Kratz, F. Albumin as a drug carrier: Design of prodrugs, drug conjugates and nanoparticles. *J. Control. Release* **2008**, *132*, 171–183. [CrossRef] [PubMed]
126. Haley, B.; Frenkel, E. Nanoparticles for drug delivery in cancer treatment. *Urol. Oncol.* **2008**, *26*, 57–64. [CrossRef]
127. Lamichhane, S.; Lee, S. Albumin nanoscience: Homing nanotechnology enabling targeted drug delivery and therapy. *Arch. Pharm. Res.* **2020**, *43*, 118–133. [CrossRef] [PubMed]
128. John, T.A.; Vogel, S.M.; Minshall, R.D.; Ridge, K.; Tiruppathi, C.; Malik, A.B. Evidence for the role of alveolar epithelial gp60 in active transalveolar albumin transport in the rat lung. *J. Physiol.* **2001**, *533*, 547–559. [CrossRef] [PubMed]
129. Maeda, H.; Bharate, G.Y.; Daruwalla, J. Polymeric drugs for efficient tumor-targeted drug delivery based on EPR-effect. *Eur. J. Pharm. Biopharm.* **2009**, *71*, 409–419. [CrossRef]
130. Fang, J.; Nakamura, H.; Maeda, H. The EPR effect: Unique features of tumor blood vessels for drug delivery, factors involved, and limitations and augmentation of the effect. *Adv. Drug Deliv. Rev.* **2011**, *63*, 136–151. [CrossRef]
131. Preeti, K.; Balaram, G.; Swati, B. Nanocarriers for cancer-targeted drug delivery. *J. Drug Target.* **2015**. [CrossRef]
132. Fang, J.; Islam, R.; Islam, W.; Yin, H.; Subr, V.; Etrych, T.; Ulbrich, K.; Maeda, H. Augmentation of EPR Effect and Efficacy of Anticancer Nanomedicine by Carbon Monoxide Generating Agents. *Pharmaceutics* **2019**, *11*, 343. [CrossRef]
133. Maeda, H. Toward a full understanding towards EPR effect in primary and metastatic tumors as well as issues related to its heterogeneity. *Adv. Drug Deliv. Rev.* **2015**. [CrossRef]
134. Pearce, A.K.; O'Reilly, R.K. Insights into Active Targeting of Nanoparticles in Drug Delivery: Advances in Clinical Studies and Design Considerations for Cancer Nanomedicine. *Bioconjug. Chem.* **2019**, *30*, 2300–2311. [CrossRef]
135. Ahmad, A.; Khan, F.; Mishra, R.K.; Khan, R. Precision Cancer Nanotherapy: Evolving Role of Multifunctional Nanoparticles for Cancer Active Targeting. *J. Med. Chem.* **2019**, *62*, 10475–10496. [CrossRef] [PubMed]
136. Ahmad, N.; Ahmad, R.; Alam, M.A.; Ahmad, F.J. Enhancement of oral bioavailability of doxorubicin through surface modified biodegradable polymeric nanoparticles. *Chem. Cent. J.* **2018**, *12*. [CrossRef] [PubMed]
137. Soma, C.; Dubernet, C.; Bentolila, D.; Benita, S.; Couvreur, P. Reversion of multidrug resistance by co-encapsulation of doxorubicin and cyclosporin A in polyalkylcyanoacrylate nanoparticles. *Biomaterials* **2002**, *21*, 1–7. [CrossRef]
138. Zhao, Y.; Cai, C.; Liu, M.; Pei, W.; Chu, X.; Zhang, H.; Wang, Z.; Han, J. An organic solvent-free technology for the fabrication of albumin-based paclitaxel nanoparticles for effective cancer therapy. *Colloids Surf. B Biointerfaces* **2019**, *183*, 110394. [CrossRef] [PubMed]
139. Çirpanli, Y.; Allard, E.; Passirani, C.; Bilensoy, E.; Lemaire, L.; Çali, S.; Benoit, J. Antitumoral activity of camptothecin-loaded nanoparticles in 9L rat glioma model. *Int. J. Pharm.* **2011**, *403*, 201–206. [CrossRef] [PubMed]
140. Guo, J.; Gao, X.; Su, L.; Xia, H.; Gu, G.; Pang, Z.; Jiang, X.; Yao, L.; Chen, J.; Chen, H. Aptamer-functionalized PEG–PLGA nanoparticles for enhanced anti-glioma drug delivery. *Biomaterials* **2011**, *32*, 8010–8020. [CrossRef]
141. Malinovskaya, Y.; Melnikov, P.; Baklaushev, V.; Gabashvili, A.; Osipova, N.; Mantrov, S.; Ermolenko, Y.; Maksimenko, O.; Gorshkola, M.; Balabanyan, V.; et al. Delivery of doxorubicin-loaded PLGA nanoparticles into U87 human glioblastoma cells. *Int. J. Pharm.* **2017**, *524*, 77–80. [CrossRef]
142. Hekmatara, T.; Bernreuther, C.; Khalansky, A.; Theisen, A.; Weissenberger, J.; Matschke, J.; Gelperina, S.; Kreuter, J.; Glatzel, M. Efficient systemic therapy of rat glioblastoma by nanoparticle-bound doxorubicin is due to antiangiogenic effects. *Clin. Neuropathol.* **2009**, *28*, 153–164. [CrossRef]
143. Khanna, V.; Kalscheuer, S.; Kirtane, A.; Zhang, W.; Panyam, J. Perlecan-targeted nanoparticles for drug delivery to triple-negative breast cancer. *Future Drug Discov.* **2019**, *1*, FDD8. [CrossRef]
144. Yuan, J.; ZhuGe, D.; Tong, M.; Lin, M.; Xu, X.; Tang, X.; Zhao, Y.; Xu, H. pH-Sensitive polymeric nanoparticles of mPEG-PLGA-PGlu with hybrid core for simultaneous encapsulation of curcumin and doxorubicin to kill the heterogeneous tumour cells in breast cancer. *Artif. Cells Nanomed. Biotechnol.* **2018**, 1–12. [CrossRef]

145. Hu, D.; Chen, L.; Qu, Y.; Peng, J.; Chu, B.; Shi, K.; Hao, Y.; Zhong, L.; Wang, M.; Qian, Z. Oxygen-generating hybrid polymeric nanoparticles with encapsulated doxorubicin and chlorin e6 for trimodal imaging-guided combined chemo-photodynamic therapy. *Theranostics* **2018**, *8*, 1558–1574. [CrossRef] [PubMed]
146. Shafiei-Irannejad, V.; Samadi, N.; Salehi, R.; Yousefi, B.; Rahimi, M.; Akbarzadeh, A.; Zarghami, N. Reversion of Multidrug Resistance by Co-Encapsulation of Doxorubicin and Metformin in Poly (lactide-co-glycolide)-d-α-tocopheryl Polyethylene Glycol 1000 Succinate Nanoparticles. *Pharm. Res.* **2018**, *35*, 119. [CrossRef]
147. Gao, J.; Liu, J.; Xie, F.; Lu, Y.; Yin, C.; Shen, X. Co-Delivery of Docetaxel and Salinomycin to Target Both Breast Cancer Cells and Stem Cells by PLGA/TPGS Nanoparticles. *Int. J. Nanomed.* **2019**, *14*, 9199–9216. [CrossRef] [PubMed]
148. Li, Z.; Chen, Q.; Qi, Y.; Liu, Z.; Hao, T.; Sun, X.; Qiao, M.; Ma, X.; Xu, T.; Zhao, X.; et al. Rational Design of Multifunctional Polymeric Nanoparticles Based on Poly(l-histidine) and d-α-Vitamin E Succinate for Reversing Tumor Multidrug Resistance. *Biomacromolecules* **2018**, *19*, 2595–2609. [CrossRef] [PubMed]
149. Zhu, D.; Tao, W.; Zhang, H.; Liu, G.; Wang, T.; Zhang, L.; Zeng, X.; Mei, L. Docetaxel (DTX)-loaded polydopamine-modified TPGS-PLA nanoparticles as a targeted drug delivery system for the treatment of liver cancer. *Acta Biomater.* **2016**, *30*, 144–154. [CrossRef] [PubMed]
150. Hu, J.; Fu, S.; Peng, Q.; Han, Y.; Xie, J.; Zan, N.; Chen, Y.; Fan, J. Paclitaxel-loaded polymeric nanoparticles combined with chronomodulated chemotherapy on lung cancer: In vitro and in vivo evaluation. *Int. J. Pharm.* **2017**, *516*, 313–322. [CrossRef]
151. Jin, M.; Jin, G.; Kang, L.; Chen, L.; Gao, Z.; Huang, W. Smart polymeric nanoparticles with pH-responsive and PEG-detachable properties for co-delivering paclitaxel and survivin siRNA to enhance antitumor outcomes. *Int. J. Nanomed.* **2018**, *13*, 2405–2423. [CrossRef]
152. Tseng, C.; Wang, T.; Dong, G.; Wu, S.H.; Young, T.; Shieh, M.; Lou, P.; Lin, F. Development of gelatin nanoparticles with biotinylated EGF conjugation for lung cancer targeting. *Biomaterials* **2018**, *28*, 3996–4005. [CrossRef]
153. Tseng, C.-L.; Wu, S.Y.-H.; Wang, W.-H.; Peng, C.-L.; Lin, F.-H.; Lin, C.; Young, T.-H.; Shieh, M.J. Targeting efficiency and biodistribution of biotinylated-EGF-conjugated gelatin nanoparticles administered via aerosol delivery in nude mice with lung cancer. *Biomaterials* **2008**, *29*, 3014–3022. [CrossRef]
154. Tseng, C.; Su, W.; Yen, K.; Yang, K.; Lin, F. The use of biotinylated-EGFmodified gelatin nanoparticle carrier to enhance cisplatin accumulation in cancerous lungs via inhalation. *Biomaterials* **2009**, *30*, 3476–3485. [CrossRef]
155. Choi, S.; Byeon, H.; Choi, J.; Thao, L.; Kim, I.; Lee, E.; Shin, B.; Lee, K.; Youn, Y. Inhalable self-assembled albumin nanoparticles for treating drug resistant lung cancer. *J. Control. Release* **2015**, *197*, 199–207. [CrossRef] [PubMed]
156. Dong, M.; Murdter, T.; Philippi, C.; Loretz, B.; Schaefer, U.; Lehr, C.; Schwab, M.; Ammon-Treiber, S. Pulmonary delivery and tissue distribution of aerosolized antisense 2′-O-Methyl RNA containing nanoplexes in the isolated perfused and ventilated rat lung. *Eur. J. Pharm. Biopharm.* **2012**, *81*, 478–485. [CrossRef] [PubMed]
157. Nafee, N.; Schneider, M.; Friebel, K.; Dong, M.; Schaefer, U.; Murdter, T.; Lehr, C. Treatment of lung cancer via telomerase inhibition: Self-assembled nanoplexes versus polymeric nanoparticles as vectors for 2′-O-Methyl-RNA. *Eur. J. Pharm. Biopharm.* **2012**, *80*, 478–489. [CrossRef] [PubMed]
158. Al-Hallak, K.M.; Azarmi, S.; Anwar-Mohamed, A.; Roa, W.H.; Löbenberg, R. Secondary cytotoxicity mediated by alveolar macrophages: A contribution to the total efficacy of nanoparticles in lung cancer therapy? *Eur. J. Pharm. Biopharm.* **2010**, *76*, 112–119. [CrossRef]
159. Zhong, Q.; da Rocha, S.R. Poly (amidoamine) Dendrimer-Doxorubicin Conjugates: In Vitro Characteristics and Pseudosolution Formulation in Pressurized Metered-Dose Inhalers. *Mol. Pharm.* **2016**, *13*, 1058–1072. [CrossRef]
160. Xie, Y.; Aillon, K.L.; Cai, S.; Christian, J.M.; Davies, N.M.; Berkland, C.J.; Forrest, M.L. Pulmonary delivery of cisplatin-hyaluronan conjugates via endotracheal instillation for the treatment of lung cancer. *Int. J. Pharm.* **2010**, *392*, 156–163. [CrossRef]
161. Gautam, A.; Densmore, C.L.; Melton, S.; Golunski, E.; Waldrep, J.C. Aerosol delivery of PEI-p53 complexes inhibits B16-F10 lung metastases through regulation of angiogenesis. *Cancer Gene Ther.* **2002**, *9*, 28–36. [CrossRef]

162. Hong, S.; Lee, J.; Jiang, H.; Kim, J.; Lee, A.; Kim, S.; Cho, C.; Cho, M. Dual expression of shAkt1 and Pdcd4 suppresses lung tumorigenesis in K-rasLA1 mice. *Anticancer Res.* **2015**, *35*, 2015–2019.
163. Kim, H.W.; Park, I.K.; Cho, C.S.; Lee, K.H.; Beck, G.R.; Colburn, N.H.; Cho, M.H. Aerosol delivery of glucosylated polyethylenimine/phosphatase and tensin homologue deleted on chromosome 10 complex suppresses Akt downstream pathways in the lung of K-ras null mice. *Cancer Res.* **2004**, *64*, 7971–7976. [CrossRef]
164. Jin, H.; Kim, T.H.; Hwang, S.K.; Chang, S.H.; Kim, H.W.; Anderson, H.K.; Lee, H.W.; Lee, K.H.; Colburn, N.H.; Yang, H.S.; et al. Aerosol delivery of urocanic acid-modified chitosan/programmed cell death 4 complex regulated apoptosis, cell cycle, and angiogenesis in lungs of K-ras null mice. *Mol. Cancer Ther.* **2006**, *5*, 1041–1049. [CrossRef]
165. Jin, H.; Xu, C.X.; Kim, H.W.; Chung, Y.S.; Shin, J.Y.; Chang, S.H.; Park, S.J.; Lee, E.S.; Hwang, S.K.; Kwon, J.T.; et al. Urocanic acid-modified chitosan-mediated PTEN delivery via aerosol suppressed lung tumorigenesis in K-ras(LA1) mice. *Cancer Gene Ther.* **2008**, *15*, 275–283. [CrossRef] [PubMed]
166. Jiang, H.L.; Hong, S.H.; Kim, Y.K.; Islam, M.A.; Kim, H.J.; Choi, Y.J.; Nah, J.W.; Lee, K.H.; Han, K.W.; Chae, C.; et al. Aerosol delivery of spermine-based poly(amino ester)/Akt1 shRNA complexes for lung cancer gene therapy. *Int. J. Pharm.* **2011**, *420*, 256–265. [CrossRef] [PubMed]
167. Kim, Y.K.; Cho, C.S.; Cho, M.H.; Jiang, H.L. Spermine-alt-poly(ethylene glycol) polyspermine as a safe and efficient aerosol gene carrier for lung cancer therapy. *J. Biomed. Mater Res. A* **2014**, *102*, 2230–2237. [CrossRef] [PubMed]
168. Jiang, H.L.; Kim, Y.K.; Arote, R.; Nah, J.W.; Cho, M.H.; Choi, Y.J.; Akaike, T.; Cho, C.S. Chitosan-graft-polyethylenimine as a gene carrier. *J. Control. Release* **2007**, *117*, 273–280. [CrossRef] [PubMed]
169. Jiang, H.L.; Xu, C.X.; Kim, Y.K.; Arote, R.; Jere, D.; Lim, H.T.; Cho, M.H.; Cho, C.S. The suppression of lung tumorigenesis by aerosol-delivered folate-chitosan-graft-polyethylenimine/Akt1 shRNA complexes through the Akt signaling pathway. *Biomaterials* **2009**, *30*, 5844–5852. [CrossRef] [PubMed]
170. Zou, Y.; Tornos, C.; Qiu, X.; Lia, M.; Perez-Soler, R. p53 Aerosol formulation with low toxicity and high efficiency for early lung cancer treatment. *Clin. Cancer Res.* **2007**, *13*, 4900–4908. [CrossRef]
171. Xu, C.X.; Jere, D.; Jin, H.; Chang, S.H.; Chung, Y.S.; Shin, J.Y.; Kim, J.E.; Park, S.J.; Lee, Y.H.; Chae, C.H.; et al. Poly(ester amine)-mediated, aerosol-delivered Akt1 small interfering RNA suppresses lung tumorigenesis. *Am. J. Respir. Crit. Care Med.* **2008**, *178*, 60–73. [CrossRef]
172. Xu, C.; Wang, P.; Zhang, J.; Tian, H.; Park, K.; Chen, X. Pulmonary Codelivery of Doxorubicin and siRNA by pH-Sensitive Nanoparticles for Therapy of Metastatic Lung Cancer. *Small* **2015**, *11*, 4321–4333. [CrossRef]
173. Xu, C.; Tian, H.; Sun, H.; Jiao, Z.; Zhang, Y.; Chen, X. A pH sensitive co-delivery system of siRNA and doxorubicin for pulmonary administration to B16F10 metastatic lung cancer. *RSC Adv.* **2015**, *5*, 103380–103385. [CrossRef]
174. Lin, G.; Zhang, H.; Huang, L. Smart polymeric nanoparticles for cancer gene delivery. *Mol. Pharm.* **2015**, *12*, 314–321. [CrossRef]
175. Santini, A.; Cammarata, S.M.; Capone, G.; Ianaro, A.; Tenore, G.C.; Pani, L.; Novellino, E. Nutraceuticals: Opening the debate for a regulatory framework. *Br. J. Clin. Pharmacol.* **2018**, *84*, 659–672. [CrossRef] [PubMed]
176. Aronson, J.K. Defining 'nutraceuticals': Neither nutritious nor pharmaceutical. *Br. J. Clin. Pharmacol.* **2017**, *83*, 8–19. [CrossRef] [PubMed]
177. Rajapakse, T.; Pringsheim, T. Nutraceuticals in Migraine: A Summary of Existing Guidelines for Use. *Headache* **2016**, *56*, 808–816. [CrossRef] [PubMed]
178. Santini, A.; Novellino, E. Nutraceuticals in hypercholesterolaemia: An overview. *Br. J. Clin. Pharmacol.* **2017**, *174*, 1450–1463. [CrossRef] [PubMed]
179. Rivellese, A.A.; Ciciola, P.; Costabile, G.; Vetrani, C.; Vitale, M. The Possible Role of Nutraceuticals in the Prevention of Cardiovascular Disease. *High Blood Press. Cardiovasc. Prev. Off. J. Ital. Soc. Hypertens.* **2019**, *26*, 101–111. [CrossRef]
180. Poli, A.; Visioli, F. Pharmacology of Nutraceuticals with Lipid Lowering Properties. *High Blood Press. Cardiovasc. Prev. Off. J. Ital. Soc. Hypertens.* **2019**, *26*, 113–118. [CrossRef]

181. Borghi, C.; Cicero, A.F.G. Nutraceuticals with a clinically detectable blood pressure-lowering effect: A review of available randomized clinical trials and their meta-analyses. *Br. J. Clin. Pharmacol.* **2017**, *83*, 163–171. [CrossRef]
182. Brown, T.J.; Brainard, J.; Song, F.; Wang, X.; Abdelhamid, A.; Hooper, L.; Group, P. Omega-3, omega-6, and total dietary polyunsaturated fat for prevention and treatment of type 2 diabetes mellitus: Systematic review and meta-analysis of randomised controlled trials. *BMJ* **2019**, *366*, l4697. [CrossRef]
183. Jovanovski, E.; Khayyat, R.; Zurbau, A.; Komishon, A.; Mazhar, N.; Sievenpiper, J.L.; Blanco Mejia, S.; Ho, H.V.T.; Li, D.; Jenkins, A.L.; et al. Should Viscous Fiber Supplements Be Considered in Diabetes Control? Results from a Systematic Review and Meta-analysis of Randomized Controlled Trials. *Diabetes Care* **2019**, *42*, 755–766. [CrossRef]
184. Varela-López, A.; Navarro-Hortal, M.D.; Giampieri, F.; Bullón, P.; Battino, M.; Quiles, J.L. Nutraceuticals in Periodontal Health: A Systematic Review on the Role of Vitamins in Periodontal Health Maintenance. *Molecules* **2018**, *23*, 1226. [CrossRef]
185. Orr, S.L. The Evidence for the Role of Nutraceuticals in the Management of Pediatric Migraine: A Review. *Curr. Pain Headache Rep.* **2018**, *22*, 37. [CrossRef] [PubMed]
186. Wang, A.; Leong, D.J.; Cardoso, L.; Sun, H.B. Nutraceuticals and osteoarthritis pain. *Pharmacol. Ther.* **2018**, *187*, 167–179. [CrossRef] [PubMed]
187. Jones, D.; Caballero, S.; Davidov-Pardo, G. Bioavailability of nanotechnology-based bioactives and nutraceuticals. *Adv. Food Nutr. Res.* **2019**, *88*, 235–273. [CrossRef] [PubMed]
188. Liu, C.-S.; Zheng, Y.-R.; Zhang, Y.-F.; Long, X.-Y. Research progress on berberine with a special focus on its oral bioavailability. *Fitoterapia* **2016**, *109*, 274–282. [CrossRef]
189. Sharma, R.A.; Euden, S.A.; Platton, S.L.; Cooke, D.N.; Shafayat, A.; Hewitt, H.R.; Marczylo, T.H.; Morgan, B.; Hemingway, D.; Plummer, S.M.; et al. Phase I clinical trial of oral curcumin: Biomarkers of systemic activity and compliance. *Clin. Cancer Res. Off. J. Am. Assoc. Cancer Res.* **2004**, *10*, 6847–6854. [CrossRef]
190. Pollack, R.M.; Barzilai, N.; Anghel, V.; Kulkarni, A.S.; Golden, A.; O'Broin, P.; Sinclair, D.A.; Bonkowski, M.S.; Coleville, A.J.; Powell, D.; et al. Resveratrol Improves Vascular Function and Mitochondrial Number but Not Glucose Metabolism in Older Adults. *J. Gerontol. Ser. A Biol. Sci. Med Sci.* **2017**, *72*, 1703–1709. [CrossRef] [PubMed]
191. Fan, D.; Wu, X.; Dong, W.; Sun, W.; Li, J.; Tang, X. Enhancement by sodium caprate and sodium deoxycholate of the gastrointestinal absorption of berberine chloride in rats. *Drug Dev. Ind. Pharm.* **2013**, *39*, 1447–1456. [CrossRef] [PubMed]
192. Kawabata, K.; Yoshioka, Y.; Terao, J. Role of Intestinal Microbiota in the Bioavailability and Physiological Functions of Dietary Polyphenols. *Molecules* **2019**, *24*, 370. [CrossRef] [PubMed]
193. Piskula, M.K.; Murota, K.; Terao, J. Bioavailability of flavonols and flavones. In *Flavonoids and Related Compounds: Bioavailability and Function*; CRC Press: Boca Raton, FL, USA, 2012; pp. 93–107.
194. Rich, G.T.; Buchweitz, M.; Winterbone, M.S.; Kroon, P.A.; Wilde, P.J. Towards an Understanding of the Low Bioavailability of Quercetin: A Study of Its Interaction with Intestinal Lipids. *Nutrients* **2017**, *9*, 111. [CrossRef] [PubMed]
195. Hisanaga, A.; Mukai, R.; Sakao, K.; Terao, J.; Hou, D.-X. Anti-inflammatory effects and molecular mechanisms of 8-prenyl quercetin. *Mol. Nutr. Food Res.* **2016**, *60*, 1020–1032. [CrossRef] [PubMed]
196. Barahona, M.J.; Baratta, V.; Ollodart, J.; Mulligan, D.; Geibel, J.P. Design and implementation of novel nutraceuticals and derivatives for treating intestinal disorders. *Future Med. Chem.* **2019**, *11*, 847–855. [CrossRef] [PubMed]
197. Calvo-Castro, L.A.; Schiborr, C.; David, F.; Ehrt, H.; Voggel, J.; Sus, N.; Behnam, D.; Bosy-Westphal, A.; Frank, J. The Oral Bioavailability of Trans-Resveratrol from a Grapevine-Shoot Extract in Healthy Humans is Significantly Increased by Micellar Solubilization. *Mol. Nutr. Food Res.* **2018**, *62*, e1701057. [CrossRef] [PubMed]
198. Xiong, W.; Ren, C.; Li, J.; Li, B. Enhancing the photostability and bioaccessibility of resveratrol using ovalbumin-carboxymethylcellulose nanocomplexes and nanoparticles. *Food Funct.* **2018**, *9*, 3788–3797. [CrossRef] [PubMed]
199. Jaisamut, P.; Wiwattanawongsa, K.; Wiwattanapatapee, R. A Novel Self-Microemulsifying System for the Simultaneous Delivery and Enhanced Oral Absorption of Curcumin and Resveratrol. *Planta Med.* **2017**, *83*, 461–467. [CrossRef] [PubMed]

200. Yadav, P.; Bandyopadhyay, A.; Chakraborty, A.; Sarkar, K. Enhancement of anticancer activity and drug delivery of chitosan-curcumin nanoparticle via molecular docking and simulation analysis. *Carbohydr. Polym.* **2018**, *182*, 188–198. [CrossRef]
201. Jaiswal, S.; Mishra, P. Co-delivery of curcumin and serratiopeptidase in HeLa and MCF-7 cells through nanoparticles show improved anti-cancer activity. *Mater. Sci. Eng. C Mater. Biol. Appl.* **2018**, *92*, 673–684. [CrossRef]
202. Liu, L.; Xiong, X.; Shen, M.; Ru, D.; Gao, P.; Zhang, X.; Huang, C.; Sun, Y.; Li, H.; Duan, Y. Co-Delivery of Triptolide and Curcumin for Ovarian Cancer Targeting Therapy via mPEG-DPPE/CaP Nanoparticle. *J. Biomed. Nanotechnol.* **2018**, *14*, 1761–1772. [CrossRef]
203. Xiao, B.; Zhang, Z.; Viennois, E.; Kang, Y.; Zhang, M.; Han, M.K.; Chen, J.; Merlin, D. Combination Therapy for Ulcerative Colitis: Orally Targeted Nanoparticles Prevent Mucosal Damage and Relieve Inflammation. *Theranostics* **2016**, *6*, 2250–2266. [CrossRef]
204. Gugulothu, D.; Kulkarni, A.; Patravale, V.; Dandekar, P. pH-sensitive nanoparticles of curcumin-celecoxib combination: Evaluating drug synergy in ulcerative colitis model. *J. Pharm. Sci.* **2014**, *103*, 687–696. [CrossRef]
205. Niazvand, F.; Khorsandi, L.; Abbaspour, M.; Orazizadeh, M.; Varaa, N.; Maghzi, M.; Ahmadi, K. Curcumin-loaded poly lactic-co-glycolic acid nanoparticles effects on mono-iodoacetate-induced osteoarthritis in rats. *Vet. Res. Forum Int. Q. J.* **2017**, *8*, 155–161.
206. Ohno, M.; Nishida, A.; Sugitani, Y.; Nishino, K.; Inatomi, O.; Sugimoto, M.; Kawahara, M.; Andoh, A. Nanoparticle curcumin ameliorates experimental colitis via modulation of gut microbiota and induction of regulatory T cells. *PLoS ONE* **2017**, *12*, e0185999. [CrossRef]
207. Natesan, S.; Pandian, S.; Ponnusamy, C.; Palanichamy, R.; Muthusamy, S.; Kandasamy, R. Co-encapsulated resveratrol and quercetin in chitosan and peg modified chitosan nanoparticles: For efficient intra ocular pressure reduction. *Int. J. Biol. Macromol.* **2017**, *104*, 1837–1845. [CrossRef] [PubMed]
208. Song, Z.; Shi, Y.; Han, Q.; Dai, G. Endothelial growth factor receptor-targeted and reactive oxygen species-responsive lung cancer therapy by docetaxel and resveratrol encapsulated lipid-polymer hybrid nanoparticles. *Biomed. Pharmacother.* **2018**, *105*, 18–26. [CrossRef] [PubMed]
209. Siu, F.Y.; Ye, S.; Lin, H.; Li, S. Galactosylated PLGA nanoparticles for the oral delivery of resveratrol: Enhanced bioavailability and in vitro anti-inflammatory activity. *Int. J. Nanomed.* **2018**, *13*, 4133–4144. [CrossRef] [PubMed]
210. Soukoulis, C.; Bohn, T. A comprehensive overview on the micro- and nano-technological encapsulation advances for enhancing the chemical stability and bioavailability of carotenoids. *Crit. Rev. Food Sci. Nutr.* **2018**, *58*, 1–36. [CrossRef]
211. Yi, J.; Lam, T.I.; Yokoyama, W.; Cheng, L.W.; Zhong, F. Beta-carotene encapsulated in food protein nanoparticles reduces peroxyl radical oxidation in Caco-2 cells. *Food Hydrocoll.* **2015**, *43*, 31–40. [CrossRef]
212. Massounga Bora, A.F.; Ma, S.; Li, X.; Liu, L. Application of microencapsulation for the safe delivery of green tea polyphenols in food systems: Review and recent advances. *Food Res. Int.* **2018**, *105*, 241–249. [CrossRef]
213. Mazzanti, G.; Menniti-Ippolito, F.; Moro, P.A.; Cassetti, F.; Raschetti, R.; Santuccio, C.; Mastrangelo, S. Hepatotoxicity from green tea: A review of the literature and two unpublished cases. *Eur. J. Clin. Pharmacol.* **2009**, *65*, 331–341. [CrossRef] [PubMed]
214. Nowak, A.; Sójka, M.; Klewicka, E.; Lipińska, L.; Klewicki, R.; Kołodziejczyk, K. Ellagitannins from Rubus idaeus L. Exert Geno- and Cytotoxic Effects against Human Colon Adenocarcinoma Cell Line Caco-2. *J. Agric. food Chem.* **2017**, *65*, 2947–2955. [CrossRef]
215. Gupta, R.C.; Srivastava, A.; Lall, R. Toxicity Potential of Nutraceuticals. *Methods Mol. Biol.* **2018**, *1800*, 367–394. [CrossRef]
216. Zhang, C.; Li, Y.; Liu, L.; Gong, Y.; Xie, Y.; Cao, Y. Chemical Structures of Polyphenols That Critically Influence the Toxicity of ZnO Nanoparticles. *J. Agric. Food Chem.* **2018**, *66*, 1714–1722. [CrossRef]
217. Wu, W.; Zhang, R.; McClements, D.J.; Chefetz, B.; Polubesova, T.; Xing, B. Transformation and Speciation Analysis of Silver Nanoparticles of Dietary Supplement in Simulated Human Gastrointestinal Tract. *Environ. Sci. Technol.* **2018**, *52*, 8792–8800. [CrossRef]
218. Vozza, G.; Khalid, M.; Byrne, H.J.; Ryan, S.M.; Frias, J.M. Nutraceutical formulation, characterisation, and in-vitro evaluation of methylselenocysteine and selenocystine using food derived chitosan:zein nanoparticles. *Food Res. Int.* **2019**, *120*, 295–304. [CrossRef]

219. Alphandéry, E.; Grand-Dewyse, P.; Lefèvre, R.; Mandawala, C.; Durand-Dubief, M. Cancer therapy using nanoformulated substances: Scientific, regulatory and financial aspects. *Expert Rev. Anticancer Ther.* **2015**, *15*, 1233–1255. [CrossRef] [PubMed]
220. Mura, S.; Fattal, E.; Nicolas, J. From poly(alkyl cyanoacrylate) to squalene as core material for the design of nanomedicines. *J. Drug Target.* **2019**, *27*, 470–501. [CrossRef] [PubMed]
221. Ledford, H. Bankruptcy filing worries developers of nanoparticle cancer drugs. *Nature* **2016**, *533*, 304–305. [CrossRef] [PubMed]
222. Writer, S. Onxeo's liver cancer drug Livatag fails in phase 3 trial. In *Pharmaceutical Business Review*; 2017; Volume 2017, Available online: https://pharmaceutical-business-review.com/ (accessed on 13 June 2020).
223. Rodríguez Fernández, C. French Nanoparticle Therapy Fails to Improv Liver Cancer Treatment in Phase III. Available online: https://www.labiotech.eu/medical/onxeo-livatag-nanoparticle-cancer/ (accessed on 13 June 2020).

nanomaterials

MDPI

Article

Colorimetric, Naked-Eye Detection of Lysozyme in Human Urine with Gold Nanoparticles

Paula M. Castillo, Francisco J. Fernández-Acejo, Jose M. Carnerero, Rafael Prado-Gotor * and Aila Jimenez-Ruiz *

Department of Physical Chemistry, University of Seville, 41012 Seville, Spain; pcastillo@us.es (P.M.C.); franfdez24@gmail.com (F.J.F.-A.); jcarnerero2@us.es (J.M.C.)
* Correspondence: pradogotor@us.es (R.P.-G.); ajimenez28@us.es (A.J.-R.)

Abstract: The stabilizing effect of lysozymes to salt addition over a gold colloid are exploited in order to detect lysozymes in human urine samples. The present research is aimed at the development of a fast, naked-eye detection test for urinary lysozymuria, in which direct comparison with a colorimetric reference, allows for the immediate determination of positive/negative cases. CIEL*a*b* parameters were obtained from sample absorbance measurements, and their color difference with respect to a fixed reference point was measured by calculating the ΔE_{76} parameter, which is a measure of how well the colors can be distinguished by an untrained observer. Results show that a simple and quick test can reliably, in less than 15 min, give a positive colorimetric response in the naked eye for concentrations of a urinary lysozyme over 57.2 μg/mL. This concentration is well within the limits of that observed for leukemia-associated lysozymurias, among other disorders.

Keywords: gold nanoparticles; lysozyme; urine; colorimetry; CIELab; naked-eye detection; lysozymuria; proteinuria

Citation: Castillo, P.M.; Fernández-Acejo, F.J.; Carnerero, J.M.; Prado-Gotor, R.; Jimenez-Ruiz, A. Colorimetric, Naked-Eye Detection of Lysozyme in Human Urine with Gold Nanoparticles. *Nanomaterials* **2021**, *11*, 612. https://doi.org/10.3390/nano11030612

Academic Editors: Serge Cosnier and Daniela Iannazzo

Received: 28 January 2021
Accepted: 24 February 2021
Published: 1 March 2021

1. Introduction

Lysozyme, which is a cationic protein at physiological pH, can be eliminated through the urine because of its low molecular weight. Its charge makes it stand out among most proteins in the urine (such as albumin and globulins), which have a negative or virtually neutral charge at a biological pH [1,2]. However, as a general rule, proteins are only found in trace amounts in human urine samples obtained from healthy individuals. Some medical conditions involving either renal disfunction or the production of excess lysozyme can dramatically elevate lysozyme levels in urine (lysozymuria) [3,4]. Lysozymurias are associated with various renal disorders, such as hypokalemia, extrarenal infections, or a nephrotic syndrome [5], and have also been determined to be a distinct symptom of monocytic and myelomonocytic leukemia (both subtypes of acute myeloid leukemia) [3,4]. In the latter cases, lysozyme levels found in urine have been found to be abnormally high when compared to other disorders. Lysozymuria detection plays a key role in early diagnose of monocytic and myelomonocytic leukemias.

Noble metal nanoparticles have acquired great importance in the field of biomolecule detection due to their optical properties. The oscillation of the electronic cloud with the electric component of the light causes a strong band of absorption in the visible region, whose location (and, therefore, the color of the sample) depends on the colloid intrinsic properties and its environment [6]. In this way, when these nanoparticles interact with the biomolecule to be detected, a change of color takes place in the solution [7,8]. Specifically, gold nanoparticles (AuNPs) have been extensively used to the detection of a varied and wide variety of compounds, from proteins to DNA [9–13]. The procedure is quite simple. The interaction between the analyte and the AuNPs induces approximation (aggregation) of the nanoparticles, leading to nanoparticle plasmon coupling and a change of color from red to blue in the solution. Reverse procedures are also possible with analytes being employed to either prevent aggregation, or further the distance between aggregated particles, causing a blue-to-red color shift [14]. The former is the case for lysozymes, as shown in Scheme 1.

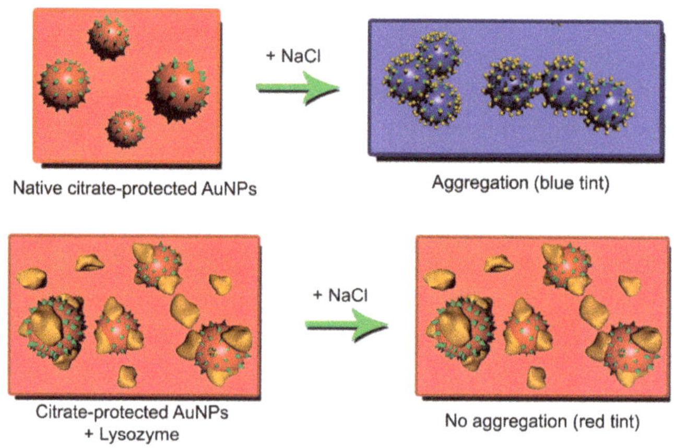

Scheme 1. Protective effect of lysozyme upon NaCl addition over citrate-capped gold nanoparticles.

Along those guidelines, multiple strategies have been carried out, from the simplest to very complex ones, in order to detect a myriad of proteins and other biological molecules. For example, cholesterol has been detected with AuNPs functionalized with cholesterol oxidase, interacting with the analyte, who makes junction points between the particles and causes the previously mentioned chromatic changes (inducing shifts of the absorbance plasmon band of c.a. 35 nm) [15]. Lin and coworkers have designed a system where the presence of the protein target (highly effective to VEGF, vascular endothelial growth factor) catalyze the formation of double chain DNA dendrimers, which cannot protect AuNPs of the aggregation induced by salt addition [16]. A similar procedure was employed by He and colleagues, who detected the AFP (Human α-fetoprotein) using the rolling circle amplification reaction as a cornerstone. The presence of the target protein generates oligonucleotides that do not prevent the aggregation caused by chloride sodium [13]. Proteins can interact directly with the nanoparticle surface, even accumulating in several layers and causing (small or big) changes in the absorption band [17].

A good number of studies have developed protocols for the detection of a lysozyme using gold nanoparticles with some of them in biological media [18–20]. Wang et al. carried out urinary lysozyme determination by employing gold nanoparticles in a really notable work that, nevertheless, employs resonance light scattering spectra for detection. Wang et al. also use lysozyme as a direct nanoparticle aggregator, so their positive response corresponds to a blue tint while a negative response gives a red tint. This method gives a lower detection threshold as lysozyme acts as an aggregating agent even when present in very small quantities [21], but it opens the possibility for samples to turn back to red when high quantities of lysozyme are present, as lysozyme presents a marked protective effect over nanoparticle aggregation [21]. Detection by aggregation is also highly vulnerable to any other interferents that may cause particles to agglomerate, such as cationic compounds. They do compensate for this by doing a very comprehensive work on pre-processing urine samples, including a protocol for the elimination of human serum albumin (HSA). This is a protein interferent that may appear alongside lysozyme in renal-damage-induced proteinuria [22]. However, their work is carried out by employing a complex detection technique that, while not as vulnerable as naked eye testing for all those factors, is also expensive and may not be available at all at testing locations.

Fei Fu et al. [19] do detect lysozyme in human blood samples by using plasmon resonance light-scattering of gold nanoparticles, but they need to consider the use of a peptidoglycan to bind to lysozyme, and they base their work on a luminescence response, involving complex measurements and equipment. Lihua Lu et al. [23] employ a novel Ir (III) complex to generate a strong luminescence response by means of a duplex DNA with a TBAG-quadruplex tail for the detection of lysozyme. The protein induces duplex dissociation of a complex of Ir (III) to generate the luminescence response.

Other authors, such as Jing Luen Wai and Siu Yee New [20], have used non-citrate AuNPs, specifically cysteamine-stabilised AuNPs (cysAuNPs). The great advantage is that these nanoparticles can directly interact with DNA with an anionic charge, without the need for an inert electrolyte. However, in addition to functionalizing gold, they work with lysozyme-binding aptamer (LBA) and their study only refers to aqueous media. There are other, very recent studies to detect lysozyme, but all of them either employ aptamers or work with non-biological samples, such as wine samples [24,25]. Other authors use nanorods [26] or carbon nanotubes, which are also functionalized with aptamers [27]. Lixiang Zuo does lysozyme detection in urine samples, but their method involves preparation of Mn-doped ZnS quantum dots [28] as does the one employed by Zhenli Qiu and coworkers for spiked serum samples [29].

Since the aim of our work is to develop a cheap, easy to use, naked-eye lysozyme sensor that can be employed without the need for expensive equipment or training, we have employed CIEL*a*b* colorimetric parameters for the determination of the quality of our results. The CIEL*a*b* colorimetric system is an absolute color coordinate system that is based on the theory of opposite colors developed by Schrödinger [30]. That is, red and green cannot be perceived at the same time, and neither can blue and yellow due to them generating eye responses that are opposite of each other [31]. The system is defined around an illuminant (which emulates external light under controlled circumstances) and an observer (which emulates human eye structure) function [32]. With those two conditions, a white color point is generated as the zero, and color intensities are mapped alongside three axes: L*, which corresponds to the luminance of the system, a*, which represents red against green tint, and b*, which does the same to yellow versus blue (Scheme 2).

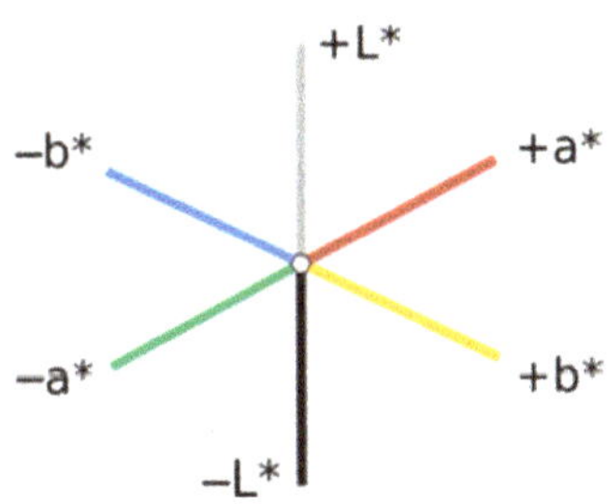

Scheme 2. Schematic representation of the three CIEL*a*b* color system axes.

CIEL*a*b* color coordinates do not depend on the device employed for reproduction (as do other systems, such as RGB for digital color or CMYK for printing), but, instead, univocally and universally define a specific tone. They are also useful for the determination of color differences, as the distance between two color points (ΔE) expressed in the L*a*b* reference system determine how different they are, or, in other words, if they will be perceived as different tones by an observer. The minimum distance between two different colors and tones of the same color is known, and has been extensively studied [33]. A value of the ΔE parameter equal to or over 2.3 for two colorimetric points is known as the Just Noticeable Difference (JND) and is considered the minimum distance needed for those two points to be considered distinguishable, under good conditions, by a human observer at a naked eye.

There is precedent for the use of CIEL*a*b colorimetric parameters for the analysis of the interaction of gold nanoparticles with a ligand [34]. However, the present work makes use of CIEL*a*b* parameters and their difference to determine if naked-eye test results will be read as different (positive response) or not (negative response) by a human observer during a field test, but the use of a colorimeter is not a requisite to carry out the testing. In this field, CIEL*a*b* parameter quantification has most notably been employed such as by Mbambo et al. in 2019 [35] who used color reproduction to create a digital color scale for the development of a salinity test for saline and estuarine water. Similar tests employing RGB colors for digital simulation on a smartphone have also been developed recently for the detection of sibutramine on food products [36].

In this work, we present a direct, simple, and novel method that allows for naked-eye detection of micromolar range concentrations of lysozyme in urine. To this end, non-functionalized, anionic gold nanoparticles have been used in order to exploit both the previously mentioned strong positive charge of lysozyme and the extraordinary optical properties of colloidal gold. The strong blue color of aggregated gold nanoparticles has been taken as the reference point. NaCl has been used as the agglomerating agent, and sodium citrate has been added after synthesis in order to act as a stabilizer, helping to reach lower detection limits. Nanoparticle size, concentration of all reactants involved, and addition order have been carefully optimized in order to develop a stable, solid, and reliable system with clearly distinguishable positive (in the presence of lysozyme) and negative signals.

2. Materials and Methods

2.1. Nanoparticle Synthesis and Stabilization

Spherical citrate-capped gold nanoparticles were synthesized by a modified Turkevich method, involving direct reduction of $HAuCl_4$ salts (Sigma-Aldrich, Darmstadt, Germany, ref. number 520918) with sodium citrate (Riedel-de Haën, Honeywell International, Charlotte, NC, USA, ref. number 32320) at 95 °C with magnetical stirring [37]. The resulting synthesis was characterized by transmission electron microscopy (TEM) and the images obtained were analyzed using ImageJ software. For preliminary tests, mean particle diameter was found to be 14.1 ± 0.9 nm, with a circularity over 92% for all cases, and the final nanoparticle concentration in the synthesis was estimated at 4.1×10^{-9} M. Working tests were carried out with a synthesis with a mean particle diameter of 15.0 ± 1.4 nm, and a concentration of 3.6×10^{-9} M. In both cases, nanoparticle synthesis was found to be monodisperse (less than a 15% size dispersion; see Figure 1).

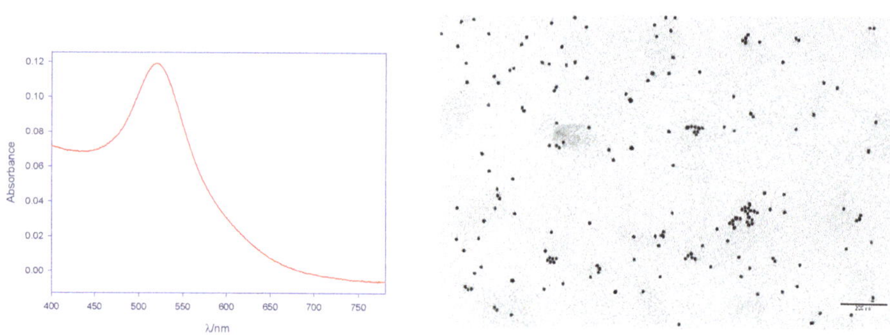

(a) (b)

Figure 1. (**a**) Absorbance spectrum and (**b**) TEM imaging of the nanoparticle synthesis employed for working tests. Scale bar corresponds to 200 nm.

Preliminary experiments with urine samples showed that the post-synthesis addition of sodium citrate greatly increased nanoparticle stability and allowed for a clearer colorimetric signal and a lower detection limit. The optimum final citrate concentration was found to be 2.3×10^{-3} M. Higher concentrations of added sodium citrate were found to have the opposite effect due to the influence of the increase in the medium ionic strength being greater than that of NP stabilization by citrate adsorption. On the other hand, when citrate concentration was increased during the synthetic procedure, nanoparticle precipitation was observed. For this reason, an additional step involving sodium citrate addition was incorporated into all sample preparation protocols, as described below.

2.2. *Preliminary Testing*

2.2.1. Sample Collection and Processing for Preliminary Testing

Urine samples from healthy subjects (labelled 1 to 3) were collected and known concentrations of lysozyme were added to each sample. Concentrations of lysozyme in the as-prepared urine solutions accounted for urinary lysozyme concentrations ranging from 10^{-5} M (143.1 µg/mL) to 5×10^{-8} M (0.72 µg/mL) and a zero sample (urine, no lysozyme). A 50% predilution (1 mL sample + 1 mL water) was then carried out in order to minimize urine color interference.

2.2.2. Preliminary Testing Protocols

Colorimetric analysis was carried out by mixing, in this order, 200 µL AuNPs + 900 µL deionized water + 200 µL sodium citrate + 200 µL prediluted urine samples + 500 µL NaCl. Final concentrations were [AuNPs] = 3.7×10^{-10} M, [Citrate] = 2.3×10^{-3} M, [NaCl] = 0.05 M. Working solutions were completely stirred and left to react for 5 min for color stabilization. In addition to the test solutions, a sample was prepared containing no added lysozyme (AuNPs + water + citrate + urine + NaCl, zero sample) and another without urine or lysozyme (AuNPs + water + citrate + NaCl, blank sample). The reproducibility of the method was evaluated by measuring three separate solutions prepared from the same initial sample (labeled as samples 1.1 to 1.3).

2.3. *Working Protocol*

2.3.1. Urine Collection and Processing

Ten urine samples from healthy subjects (labelled A to J) were collected and known concentrations of lysozyme were added to each sample. Preliminary test results allowed for the first approximation to the positive response threshold, and, therefore, lysozyme concentrations were updated. Some concentration points that were observed to be too low for detection were removed, and more points were added in the turning zone in order to better pinpoint the detection threshold. Concentrations of lysozyme in the as-prepared urine solution now accounted for urinary lysozyme concentrations ranging from 143.1 µg/mL to 7.15 µg/mL, plus a zero sample with no added lysozyme.

As per the preliminary test results, urinary salt concentration was considered to be a possible source of interference in which one where the proposed 50% urine dilution employed during preliminary sample preparation failed to properly address. Pre-treatment protocols were then updated as follows. The lysozyme was added to native urine samples in a first step 1:2 dilution. Then, the refraction index of all spiked urine samples and that of water was measured at room temperature (between 22.8 and 23.5 °C) in an Abbe WYA-1S refractometer, and a ratio $r_i = r_{sample}/r_{water}$ was then calculated. Deionized Milli-Q water was added to the samples until $r_i < 1.002$. Those diluted urine samples were then employed for sample preparation.

2.3.2. Sample Preparation

In light of the preliminary tests' results, a few modifications were made in order to develop the final working protocol. In this order, 400 µL AuNPs (1.9×10^{-8} M) + 500 µL deionized water + 400 µL sodium citrate (1.16×10^{-3} M) + 200 µL prediluted urine samples were mixed. A blank containing no urine, which was replaced by the equivalent volume of deionized water, was also prepared in the same way for each sample. A 1 M NaCl solution was then added drop-by-drop, while stirring, to this urine-free blank until a blue color shift was observed. The same volume of NaCl was then added to the sample preparations. This step allowed us to pinpoint the exact quantity of salt needed to change the color of a given gold preparation, eliminating the possibility of a false positive caused by a deficit of NaCl.

2.4. Sample Measurement and Obtainment of CIEL*a*b* Parameters

For both the preliminary and the final working protocol, samples were completely stirred and left to react for 5 min for color stabilization before measurements were done. Colorimetric analysis was carried out by measuring transmittance in a Cary 500 UV-vis spectrophotometer (Agilent, Santa Clara, CA, USA) working at room temperature. XYZ color space measurements were derived from transmittance values according to the CIE (International Commission on Illumination) standards for a D65 illuminant and 2° standard observer [38]. In order to set up quantitative guidelines for the determination of a positive response color threshold, a mathematical conversion from XYZ to L*a*b* was carried out as described by the CIE [38].

Digital (RGB) simulation of the sample colors was also carried out. Conversion from the XYZ data to RGB was done in accordance with the equations found in https://www.easyrgb.com/en/math.php (accessed on 3 March 2014). Those colors are reproduced in this paper to graphically illustrate the results of our analysis, but were not employed for mathematical calculations.

The ΔE parameter, which evaluates a color difference between a sample and a reference color, was calculated from CIEL*a*b* parameters for all cases [39]. It is important to note that ΔE values were calculated in two different ways for the preliminary tests and for the final ones.

- Preliminary tests: referenced to L*a*b* parameters of a blank sample, which does contain aggregated AuNPs without urine or lysozyme and, therefore, can be employed as a neutral reference point. However, in doing so, it was observed that zero-lysozyme samples could also show subtle color differences from the urine-free blank.
- Working tests: referenced to the mean L*a*b* parameters obtained for all zero (lysozyme-free) urine samples. In this way, a "neutral" reference that was not directly related to any of the samples was created, and residual matrix effects of urine that could induce color changes were accounted for and compensated.

For ΔE evaluation, the classical CIE76 formula (equivalent to the Euclidean distance between the reference and the sample color coordinates in a reference system) was used (Equation (1)).

$$\Delta E_{76} = \sqrt{(L^* - L^*_{Blank})^2 + (a^* - a^*_{Blank})^2 + (b^* - b^*_{Blank})^2} \quad (1)$$

3. Results

3.1. Preliminary Test Results

For the preliminary tests, upon NaCl addition, absorbance spectra were recorded, as shown in Figure 2. It is important to note that, for this representation, absorbance spectra were normalized in order to better show maxima position. Peak position showed a clear red shift (towards higher λ) at lower lysozyme concentrations, followed by blue shifting for concentrations of 71.5 μg/mL or over. The zero point of the series shows a smaller red shift than that observed for the lower lysozyme concentrations.

Naked-eye results (Figure 3) showed, in all but one of the cases, a clear red color (positive response) for urinary lysozyme concentrations of 71.5 μg/mL or higher after NaCl addition. Zero samples were blue in all cases, as were the blanks (not shown). No blue color was found in any of the samples prior to salt addition. All samples showed a color change when the lysozyme concentration was over the previously mentioned detection threshold, even though, in the outlier case, a purple tint instead of red was reached as the final point.

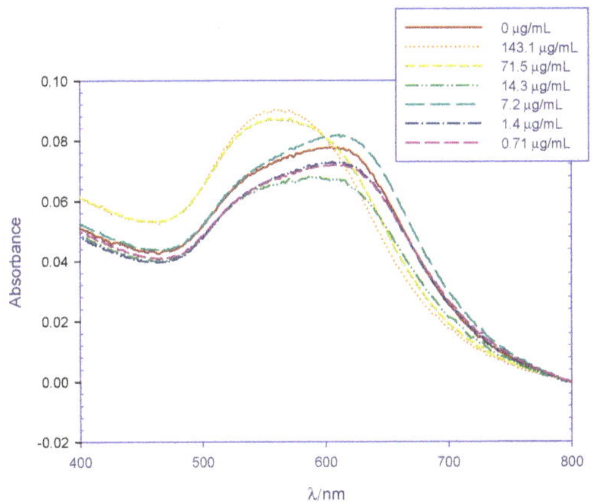

Figure 2. Normalized absorbance spectra for different concentrations of a urinary lysozyme. Series correspond to sample 1.

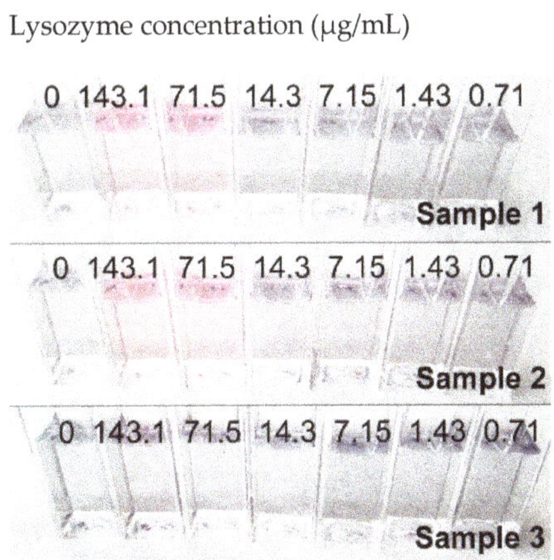

Figure 3. Preliminary naked-eye tests for lysozyme concentrations ranging from 143.1 to 0.71 µg/mL.

In order to set up quantitative guidelines for the determination of a positive response color threshold, transmittance measurements were done and CIELab parameters were derived from the results. The obtained L*a*b* values were then used for digital simulation of the sample colors, in order to allow for a clearer color observation without external illumination interference (Figure 4). Three freshly-collected samples, 1, 2, and 3 (outlier case), were analyzed, and the reproducibility of the method was evaluated by measuring three separate solutions prepared from the same initial sample (labeled as samples 1.1 to 1.3). From those measurements, standard deviation of the ΔE parameter was found to be under 10% for all cases.

In this case, ΔE values were measured in reference to the L*a*b* parameters of the blank sample, which contains aggregated AuNPs without urine or lysozyme. For ΔE evaluation, the CIE76 formula was used (see Equation (1)). Results were contrasted against those obtained from the more complex, corrected ΔE CIE94 formula [40].

$$\Delta E_{94} = \sqrt{\frac{\Delta L^{*2}}{k_L S_L} + \frac{\Delta C^{*2}}{k_C S_C} + \frac{\Delta H^{*2}}{k_H S_H}} \quad (2)$$

$$C^* = \sqrt{a^{*2} + b^{*2}} \quad (3)$$

$$\Delta C^* = C^* - C^*_{Blank} \quad (4)$$

$$\Delta H^* = \sqrt{\Delta a^{*2} + \Delta b^{*2} - \Delta C^{*2}} \quad (5)$$

$$S_L = 1 \quad (6)$$

$$S_C = 1 + 0.045\ C^*_{Blank} \quad (7)$$

$$S_H = 1 + 0.015\ C^*_{Blank} \quad (8)$$

$$k_L = k_S = k_H = 1 \quad (9)$$

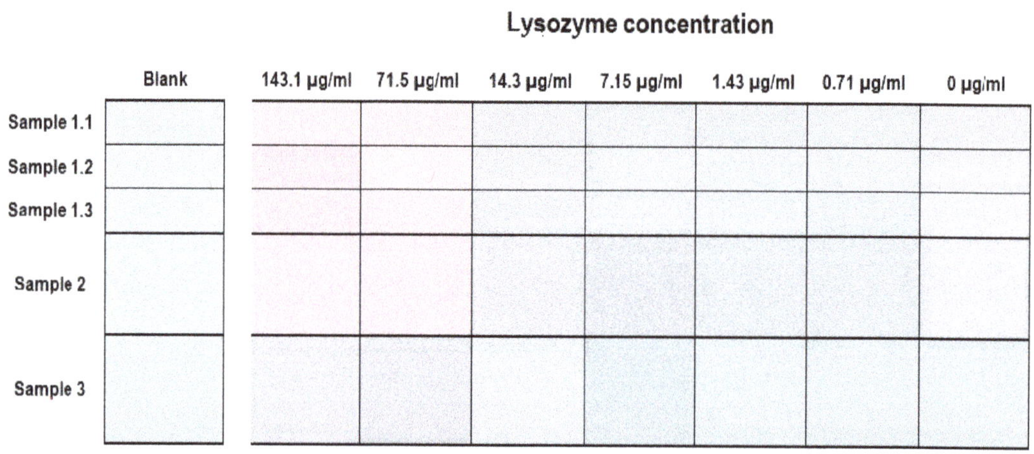

Figure 4. RGB color simulation obtained from experimental L*a*b* parameters corresponding to different lysozyme concentrations in urine. Blank samples have no urine and no lysozyme.

Despite the CIE94 system being more accurate at addressing smaller color differences [39] and non-uniform color perception [31] than the CIE76, less than a 2% difference was found in all cases between CIE76 and CIE94 values (Table 1). Since the precision gain was deemed not enough to justify the use of a more complex system, the CIE76 system was employed through the rest of our study.

Table 1. ΔE_{76} and ΔE_{94} values obtained by direct comparation of the samples with the urine-free blank. Shaded values are over the JND threshold, with darker shaded ones being over the color difference threshold.

	Sample 1.1		Sample 1.2		Sample 1.3		Sample 2		Sample 3	
[Lys]/µg·mL^{-1}	ΔE_{76}	ΔE_{94}	ΔE_{76}	ΔE_{94}	ΔE_{76}	ΔE_{94}	ΔE_{76}	ΔE_{94}	ΔE_{76}	ΔE_{94}
143.07	5.7	5.6	5.7	5.6	5.2	5.1	5.9	5.8	3.0	3.0
71.54	5.4	5.3	5.2	5.1	5.3	5.2	5.4	5.3	2.8	2.8
14.31	2.5	2.5	2.5	2.4	2.4	2.4	2.7	2.7	0.9	0.9
7.15	2.1	2.1	2.0	2.0	1.8	1.8	2.6	2.6	1.5	1.5
1.43	1.7	1.6	1.7	1.7	1.7	1.6	2.1	2.1	0.9	0.9
0.72	1.5	1.5	1.6	1.6	1.6	1.6	2.1	2.1	1.0	1.0
0	2.6	2.5	2.3	2.3	2.3	2.2	2.4	2.4	1.3	1.3

As noted before, a ΔE value of 2.3 or over is termed the JND (Just Noticeable Difference) threshold for close tones to be considered as distinguishable by an untrained observer [32]. Two tones that present a ΔE color difference under the JND are harder or impossible to distinguish from each other, while two tones whose ΔE is over the JND will be interpreted as different. The JND threshold is indicated by a dashed line in Figure 5. The most significant finding of this series is that the color difference value ΔE between the blank and zero samples in normal cases can be found in the ΔE = 2.3 to 2.6 bracket. This means that zero samples (lysozyme-free urine samples) can be distinguished from the blank sample by direct comparison, thereby, posing a risk for false positive results.

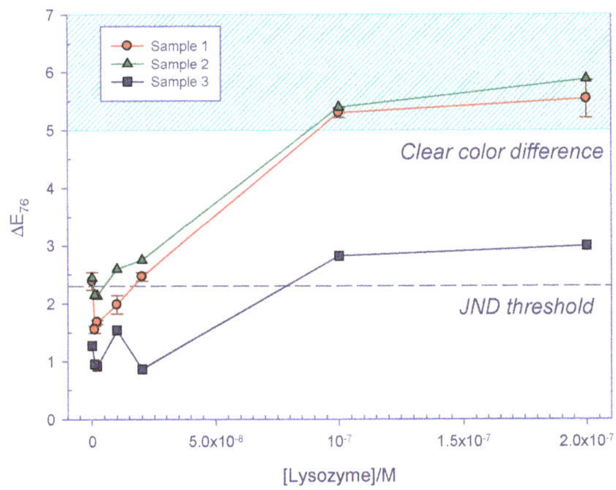

Figure 5. Sample-blank color difference ΔE_{76} evolution with increasing lysozyme concentrations. Error bars, derived from repeated measurements, are shown for Sample 1.

3.2. Working Test Results

A working test protocol differed from the preliminary one in some ways, as we tuned up our work conditions to enhance results. For a start, after the 1:2 predilution employed for native urine samples, their refractive index was also measured prior to starting the test. Pre-diluted samples were then diluted again until their $r_i = r_{sample}/r_{water}$ was under 1.002 in order to compensate for their different salt content, and then the rest of the protocol was followed.

Nanoparticle volume in preparations was raised to 400 μL from the original 200 μL, aiming for the obtention of a more intense tint without compromising detection thresholds. In addition, the "fixed volume" preparations employed through preliminary tests were changed to "dynamic volume" ones. The final volume in the detection cuvettes was not always the same, but was compensated to be as small as possible while still having the appropriate NaCl concentration to induce nanoparticle aggregation in the absence of lysozyme. To this end, each urine series was accompanied by a control sample, in which both lysozyme and urine volume were replaced by deionized water. This control sample was then added NaCl 1 M drop-by-drop until the blue tint appeared, and then the same NaCl volume was added to each point in a given series. This approach has two benefits: first, as AuNPs synthesis tend to slightly differ in concentration between batches, the minimum NaCl volume needed for aggregation may not always be the same, and, second, the risk of a "false positive" caused by a deficiency of added NaCl is also averted. Minimizing preparation volume also has the added benefit of avoiding excess AuNPs dilution, therefore, allowing for clearer red and blue tints to be observed.

As can be seen in Figure 6, the red shift of the absorbance maxima upon NaCl addition can now clearly be observed for samples 0 to 28.6 μg/mL, while upper concentrations retain the red color of unaggregated nanoparticles and appear blue-shifted in relation to the zero. No "bouncing" of the zero-lysozyme sample is observed.

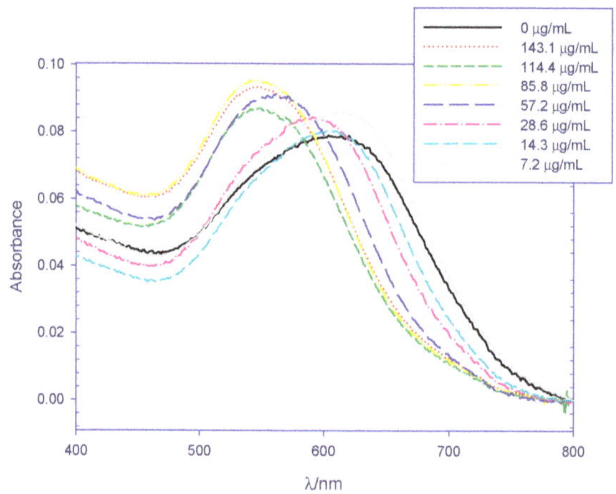

Figure 6. Normalized absorbance spectra for different concentrations of urinary lysozyme. Series correspond to sample F.

Figure 7 shows the RGB simulation of naked-eye test results. Again, for all samples, a clear red tint appears at higher lysozyme concentrations. For some of the samples, a positive response can be observed for concentrations as low as 28.6 µg/mL. It is also interesting to note that the reddish effect observed on some preliminary samples in the absence of lysozyme (0 µg/mL, see Figure 4) was corrected by the changes made to the improved protocol, minimizing the risk of a false positive.

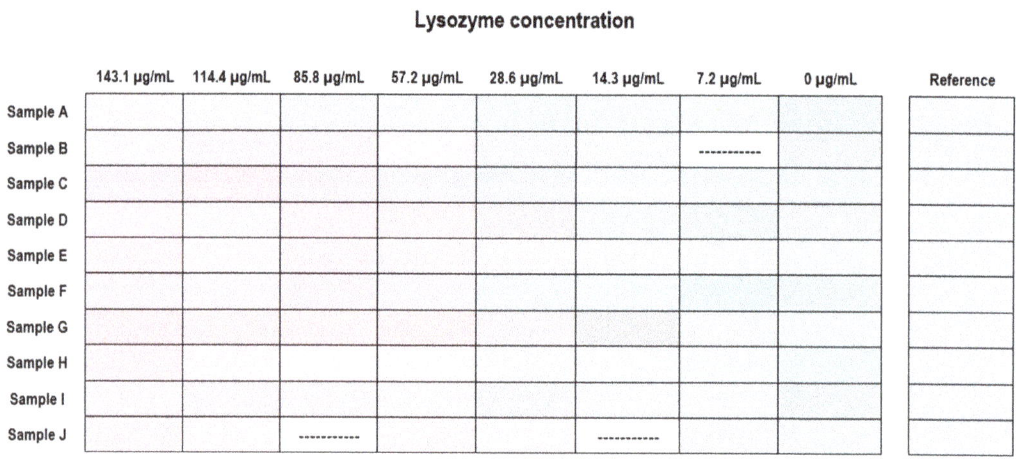

Figure 7. RGB color simulation obtained from experimental L*a*b* parameters corresponding to different lysozyme concentrations in urine.

For the experimental L*a*b* parameters of the 10 samples, ΔE values were derived. In this case, however, the reference point from which the color difference was calculated was the numerical mean of the three parameters for the zero sample of all series. In this way, the calculus compensated for the small red shift that had previously been observed for some of the zero samples. It is important to note that there was a little difference in the results when each ΔE was calculated by taking its own series zero as a reference. However, a common reference colorimetric point that can, for example, be reproduced digitally or in printing is more useful in a real setting where there is no access to a known lysozyme-free sample. The resulting L*a*b* values taken as reference were L* = 94.4, a* = −0.62, b* = −2.7.

As can be observed in Table 2 and Figure 8, for a lysozyme concentration of 114.4 μg/mL or over, all 10 samples gave positive results that could be distinguished from the reference at a naked eye. Moreover, nine out of ten cases were also over the JND threshold for lysozyme concentrations as low as 57.2 μg/mL, and four of them also gave a positive response when lysozyme concentration was halved. For the zero samples, none of them deviated so much from the reference as to constitute a false positive. This represents a marked improvement from the preliminary results in which zero samples could appear over the JND threshold.

Table 2. ΔE_{76} values obtained by direct comparison of the samples with the medium values of all zero samples. Shaded values are over the Just Noticeable Difference (JND) threshold, with darker shaded ones being over the color difference threshold.

Lysozyme (μg/mL)	A	B	C	D	E	F	G	H	I	J	Medium Value
143.1	3.1	3.5	5.2	6.1	5.3	4.0	5.9	4.9	3.7	3.9	4.7 ± 1.0
114.4	2.4	3.7	6.1	2.8	3.8	3.8	4.6	4.3	3.3	4.3	4.0 ± 1.0
85.8	1.7	4.3	4.4	6.3	4.2	4.3	5.8	4.1	3.1	-	4.4 ± 1.4
57.2	1.1	3.6	3.1	4.7	2.9	2.9	5.0	3.6	2.8	4.3	3.5 ± 1.1
28.6	0.5	2.0	1.8	3.0	2.6	1.2	3.0	2.0	1.8	2.8	2.2 ± 0.8
14.3	0.5	0.9	1.1	0.5	1.5	1.1	2.3	1.1	0.9	-	1.6 ± 0.6
7.2	1.0	-	0.9	0.6	1.1	1.0	1.0	0.9	0.8	1.1	1.3 ± 0.2
0	0.7	0.8	1.1	1.3	1.0	0.3	1.1	0.9	0.6	0.7	1.1 ± 0.3

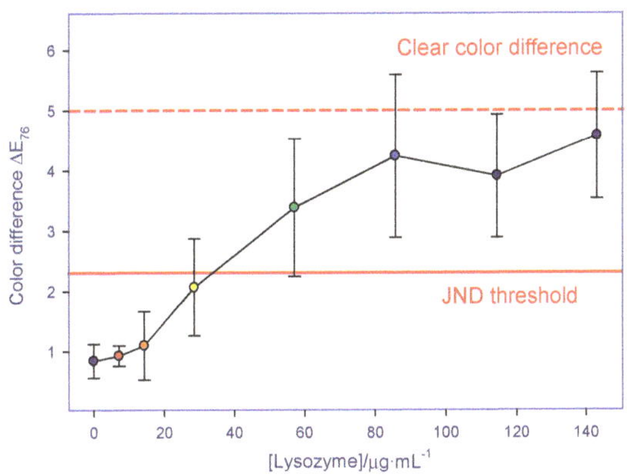

Figure 8. Mean color difference ΔE_{76} evolution for samples A–J with increasing lysozyme concentrations.

4. Discussion

4.1. Preliminary Test Results

Absorbance measurements of the lysozyme-spiked urine samples (see Figure 2) clearly showed the protective effect of lysozyme over gold nanoparticle aggregation for the more concentrated samples, which are 71.5 μg/mL and over. On the other hand, in light of those results, it becomes apparent that the multitude of processes involved with analyzing biological samples do cause a clear widening of the peaks, which is more apparent for blue-shifted results (14.3 μg/mL and under). This widening causes the exact position of the maximum to become ambiguous. This method of determining a positive response also requires spectrophotometric measurements.

For naked-eye tests, the positive signal threshold was considered to correspond to values over the JND (ΔE values of 2.3 and over), which is understood to be the smallest ΔE value needed for two tones to be perceived as different by untrained observers. As seen in Figure 4 and Table 1, results corresponding to urinary lysozyme concentrations of 71.5 µg/mL and over (clear red tint in the solution at naked eye) did reflect in all cases, but the outlier, on values of ΔE over 5, which are generally accepted to correspond to a clear color difference between two samples [33,41]. The color difference between concentrations of 71.5 and 143.1 µg/mL was almost non-existent. If all positive responses are analyzed globally, the positive response ΔE bracket extends from 5.2 to 5.7 with a mean value of ΔE = 5.5. Therefore, in the majority of the cases, the color difference observed after tuning up the system (15-nm anionic, citrate-capped AuNPs under optimum citrate and NaCl concentrations) allows for a clear distinction between a lysozyme-containing and a lysozyme-free urine sample. By direct comparison with a freshly prepared blank sample, naked-eye lysozyme detection in urine can, therefore, be carried out in a quick and easy way.

However, although the positive-blank color difference is enough to be noticeable at a glance in the outlier case 3, the difference between the positive response of this sample 3 at [lys] > 71.5 mg/mL ($\Delta E_{76} \approx 2.8$), and the 1 and 2 control samples with [lys] = 0 ($\Delta E_{76} \approx 2.4$) is not big enough to unequivocally ascertain the presence of lysozyme by the naked eye observation without either risking false positives for 1 and 2 or a false negative for 3.

The presence of a purple tint on zero-lysozyme samples can be attributed to the presence of trace amounts of other proteins. Those data are in accordance with Wang et al. [18] who reported similar observations when analyzing lysozyme-spiked urine samples through the Plasmon Resonance Light-Scattering (PRLS) technique. The higher grade of AuNPs aggregation found in sample 3, which leads to a greater blue color intensity, might be due to an abnormally high salt content in the initial sample.

4.2. *Working Test Results*

With the modified protocol, the normalized absorbance spectra of the samples (Figure 6) shows a similar behaviour to that observed for the preliminary test results. Lower lysozyme concentration samples do experience a red shifting of the absorbance spectra due to nanoparticle aggregation, while higher concentration samples remain on the green-blue absorbance zone. In this case, although the general tendency is clear, peak widening becomes even more apparent than it was during the preliminary phase of the study.

As for naked-eye testing, changes made to the preliminary testing protocol proved useful in pinpointing the detection threshold of the method. The use of a "real" reference point for the test series, obtained from the mathematical mean of the three L*a*b* parameters for the zero-lysozyme samples, meant that the possible purple tint that may appear in a lysozyme-free sample is accounted and compensated for. The reference shift also meant that less lysozyme-containing samples tested over ΔE = 5, that is, they were less distinguishable from the real reference than they were from the AuNPs + NaCl reference employed for the preliminary tests. However, all positives still tested over the JND, meaning that the color difference is still enough for the system to work.

On this ten-sample series, no false positives were observed. That is, all zero samples tested well under ΔE = 2.3 and cannot be considered as distinguishable from the reference at a naked eye. On the other hand, nine out of the ten samples showed a color change over the JND when lysozyme concentrations over 57.2 µg/mL were employed, reading as positives. For the outlier case, the color change appeared at a higher concentration threshold of 114.4 µg/mL. Concentrations under that threshold can be considered as "false negatives" in which the color test reads as a negative even in the presence of measurable lysozyme concentrations. More importantly, four out of ten samples also showed color changes over the JND for concentrations as low as 28.6 µg/mL, therefore, also testing as positives.

Based on that data, and on mean ΔE values and error margins that can be found in Figure 8, we propose a detection limit of 57.2 µg/mL for our naked-eye urinary lysozyme test.

4.3. Result Evaluation

Finally, in order to evaluate the usefulness of the proposed method, the obtained lower threshold limit needs to be put in context. Urinary lysozymuria associated with monocitic and myelomonocitic leukemia was found in a classic work to range between 25 and 420 µg/mL of lysozyme [3]. A range of 62 to 211 µg/mL has been observed for acute monocitic leukemia, and a range of 0 to 87 µg/mL has been observed for acute myelomonocitic leukemia [42]. It is important to note that urinary lysozymuria is virtually absent in other leukemia types, so its early detection may help narrow leukemia type diagnoses in a clinical setting. Lysozymuria has also been found associated with varied renal diseases, up to approximately 30 µg/mL [5]. Patients suffering from diabetic nephropathy can also show increased urinary lysozyme levels, up to 10 µg/mL [43].

As for interference testing, human albumin (HSA), which is a protein commonly found in urine [1,44], has been found to interact with gold nanoparticles [45,46]. Globulins, which is another common urinary protein family [47], have also been reported to have the same effect [45]. Those interactions, where the adsorption of proteins over the particle leads to the formation of a protein corona, may cause a small degree of protection in the presence of salt, leading to a purple color. In concrete, HSA may appear alongside or instead of lysozyme in some kinds of proteinurias, such as those induced by renal damage or diabetes [48]. Protocols for elimination of those proteins by precipitation have been proposed by Wang et al. [18]. Since the isoelectric point of those proteins is below 7 in all cases [1], they remain uncharged at a biological pH, while lysozyme presents a cationic charge. This means that selective elimination of contaminant proteins, if suspected, can be carried out relatively easily.

5. Conclusions

As stated before, a good number of studies have dealt with lysozyme detection in the last few years. However, almost all of the existing studies employed complex detection techniques that may or may not be available in a clinical setting, or made use of expensive nano-systems (for example, aptamer-based tests) in which production costs may be a limiting factor when trying to develop a first-approach commercial test to lysozyme and urinary proteinuria. Most of the existing literature also deals with methods developed either in aqueous media or non-biological samples, such as wine samples, with urine examples being scarce even in the face of the preeminence of lysozymurias and other proteinurias associated with various disorders.

In light of the data exposed throughout this manuscript, it becomes apparent that our method is able to detect urinary lysozyme concentrations within the range distinctly associated with both monocitic and myelomonocitic leukemia, among other pathologies. Both lower production costs and shorter detection times in relation to existing tests point to a method that could become a helpful aid in early detection of those pathologies, which is crucial to save lives. It is also a fast (under 10 min), easy, and inexpensive system to do first-line testing for lysozymuria, requires no specialized equipment to be carried out, and can be read by a simple, naked-eye color assessment with a reference.

Author Contributions: Conceptualization, P.M.C. and R.P.-G.; methodology, P.M.C., F.J.F.-A. and A.J.-R.; software, A.J.-R.; validation, R.P.-G. and J.M.C.; formal analysis, A.J.-R. and F.J.F.-A.; investigation, R.P.-G.; data curation, J.M.C.; writing—original draft preparation, A.J.-R.; writing—review and editing, R.P.-G. and P.M.C.; supervision, R.P.-G.; project administration, R.P.-G.; funding acquisition, R.P.-G. All authors have read and agreed to the published version of the manuscript.

Funding: This work was financed by the OTRI (2010/00000762), by the V Plan Propio Grupos Emergentes (PP2016-5937) of the University of Seville, by the Junta de Andalucía (2019/FQM-386), by the V PP USO SSGG (2019/00000570), and VI PP USO SSGG (2020/00001068 and 2020/00001073) of the University of Seville.

Institutional Review Board Statement: The study was conducted according to the guidelines of the Declaration of Helsinki, and approved by the Institutional Review Board (or Ethics Committee) of Junta de Andalucía (protocol code 0443-N-13, 15 December 2014).

Informed Consent Statement: Informed consent was obtained from all subjects involved in the study.

Data Availability Statement: The data presented in this study are available on request from the corresponding author.

Conflicts of Interest: The authors declare no conflict of interest.

References

1. Polson, A. Immune Globulin and its Use in Prophylaxis and Therapy of Infectious Diseases. *S. Afr. Med. J.* **1957**, *31*, 1242–1244. [PubMed]
2. Prin, C.; Bene, M.C.; Gobert, B.; Montagne, P.; Faure, G.C. Isoelectric restriction of human immunoglobulin isotypes. *BBA Gen. Subj.* **1995**, *1243*, 287–289. [CrossRef]
3. Osserman, E.F.; Lawlor, D.P. Serum and Urinary Lysozyme (Muramidase) in Monocytic and Monomyelocytic Leukemia. *J. Exp. Med.* **1966**, *124*, 921–952. [CrossRef]
4. Noble, R.E.; Fudenberg, H.H. Leukocyte lysozyme activity in myelocytic leukemia. *Blood* **1967**, *30*, 465–473. [CrossRef] [PubMed]
5. Harrison, J.F.; Parker, R.W.; de Silva, K.L. Lysozymuria and acute disorders of renal function. *J. Clin. Pathol.* **1973**, *26*, 278–284. [CrossRef] [PubMed]
6. Petryayeva, E.; Krull, U.J. Localized surface plasmon resonance: Nanostructures, bioassays and biosensing—A review. *Anal. Chim. Acta* **2011**, *706*, 8–24. [CrossRef] [PubMed]
7. Priyadarshini, E.; Pradhan, N. Gold nanoparticles as efficient sensors in colorimetric detection of toxic metal ions: A review. *Sens. Actuators B Chem.* **2017**, *238*, 888–902. [CrossRef]
8. Zhao, W.; Chiuman, W.; Lam, J.C.F.; Brook, M.A.; Li, Y. Simple and rapid colorimetric enzyme sensing assays using non-crosslinking gold nanoparticle aggregation. *Chem. Commun.* **2007**, *8*, 3729. [CrossRef]
9. Jazayeri, M.H.; Aghaie, T.; Avan, A.; Vatankhah, A.; Ghaffari, M.R.S. Colorimetric detection based on gold nano particles (GNPs): An easy, fast, inexpensive, low-cost and short time method in detection of analytes (protein, DNA, and ion). *Sens. Bio-Sens. Res.* **2018**, *20*, 1–8. [CrossRef]
10. Gui, G.-F.; Zhuo, Y.; Chai, Y.-Q.; Xiang, Y.; Yuan, R. A novel {ECL} biosensor for β-lactamase detection: Using RU(II) linked-ampicillin complex as the recognition element. *Biosens. Bioelectron.* **2015**, *70*, 221–225. [CrossRef] [PubMed]
11. Hazarika, P.; Ceyhan, B.; Niemeyer, C.M. Sensitive detection of proteins using difunctional DMA-gold nanoparticles. *Small* **2005**, *1*, 844–848. [CrossRef] [PubMed]
12. Zhang, Y.; Zhang, K.; Ma, H. Electrochemical DNA Biosensors Based on Gold Nanoparticles/Cysteamine/Poly(glutamic acid) Modified Electrode. *Am. J. Biomed. Sci.* **2009**, *1*, 115–125. [CrossRef]
13. Chen, C.; Luo, M.; Ye, T.; Li, N.; Ji, X.; He, Z. Sensitive colorimetric detection of protein by gold nanoparticles and rolling circle amplification. *Analyst* **2015**, *140*, 4515–4520. [CrossRef] [PubMed]
14. Zhang, Y.; McKelvie, I.D.; Cattrall, R.W.; Kolev, S.D. Colorimetric detection based on localised surface plasmon resonance of gold nanoparticles: Merits, inherent shortcomings and future prospects. *Talanta* **2016**, *152*, 410–422. [CrossRef]
15. Nirala, N.R.; Saxena, P.S.; Srivastava, A. Colorimetric detection of cholesterol based on enzyme modified gold nanoparticles. *Spectrochim. Acta Part A Mol. Biomol. Spectrosc.* **2018**, *190*, 506–512. [CrossRef]
16. Chang, C.-C.; Chen, C.-Y.; Chuang, T.-L.; Wu, T.-H.; Wei, S.-C.; Liao, H.; Lin, C.-W. Aptamer-based colorimetric detection of proteins using a branched DNA cascade amplification strategy and unmodified gold nanoparticles. *Biosens. Bioelectron.* **2016**, *78*, 200–205. [CrossRef]
17. Yang, J.A.; Johnson, B.J.; Wu, S.; Woods, W.S.; George, J.M.; Murphy, C.J. Study of wild-type α-synuclein binding and orientation on gold nanoparticles. *Langmuir* **2013**, *29*, 4603–4615. [CrossRef]
18. Wang, X.; Xu, Y.; Xu, X.; Hu, K.; Xiang, M.; Li, L.; Liu, F.; Li, N. Direct determination of urinary lysozyme using surface plasmon resonance light-scattering of gold nanoparticles. *Talanta* **2010**, *82*, 693–697. [CrossRef]
19. Fu, F.; Li, L.; Luo, Q.; Li, Q.; Guo, T.; Yu, M.; Song, Y.; Song, E. Selective and sensitive detection of lysozyme based on plasmon resonance light-scattering of hydrolyzed peptidoglycan stabilized-gold nanoparticles. *Analyst* **2018**, *143*, 1133–1140. [CrossRef]
20. Wai, J.L.; New, S.Y. Cysteamine-coated gold nanoparticles for bimodal colorimetric detection with inverse sensitivity: A proof-of-concept with lysozyme. *RSC Adv.* **2019**, *10*, 1088–1094. [CrossRef]
21. Prado-Gotor, R.; Jimenez-Ruiz, A.; Carnerero, J.M.; Grueso, E.; Villa, I. CIELab chromaticity evolution to measure the binding free energy of non-colored biomolecules to gold nanoparticles. *RSC Adv.* **2015**, *5*, 85039–85045. [CrossRef]
22. Chen, T.; Xie, N.; Viglianti, L.; Zhou, Y.; Tan, H.; Tang, B.Z.; Tang, Y. Quantitative urinalysis using aggregation-induced emission bioprobes for monitoring chronic kidney disease. *Faraday Discuss.* **2017**, *196*, 351–362. [CrossRef]
23. Lu, L.; Chan, D.S.-H.; Kwong, D.W.J.; He, H.-Z.; Leung, C.-H.; Ma, D.-L. Detection of nicking endonuclease activity using a G-quadruplex-selective luminescent switch-on probe. *Chem. Sci.* **2014**, *5*, 4561–4568. [CrossRef]

24. Mishra, R.K.; Hayat, A.; Mishra, G.K.; Catanante, G.; Sharma, V.; Marty, J.-L. A novel colorimetric competitive aptamer assay for lysozyme detection based on superparamagnetic nanobeads. *Talanta* **2017**, *165*, 436–441. [CrossRef]
25. Ortiz-Aguayo, D.; Del Valle, M. Label-Free Aptasensor for Lysozyme Detection Using Electrochemical Impedance Spectroscopy. *Sensors* **2018**, *18*, 354. [CrossRef] [PubMed]
26. Tohidi Moghadam, T.; Ranjbar, B. Heat induced aggregation of gold nanorods for rapid visual detection of lysozyme. *Talanta* **2015**, *144*, 778–787. [CrossRef]
27. Rezaei, B.; Jamei, H.R.; Ensafi, A.A. An ultrasensitive and selective electrochemical aptasensor based on rGO-MWCNTs/Chitosan/carbon quantum dot for the detection of lysozyme. *Biosens. Bioelectron.* **2018**, *115*, 37–44. [CrossRef] [PubMed]
28. Zuo, L.; Qin, G.; Lan, Y.; Wei, Y.; Dong, C. A turn-on phosphorescence aptasensor for ultrasensitive detection of lysozyme in humoral samples. *Sens. Actuators B Chem.* **2019**, *289*, 100–105. [CrossRef]
29. Qiu, Z.; Shu, J.; He, Y.; Lin, Z.; Zhang, K.; Lv, S.; Tang, D. CdTe/CdSe quantum dot-based fluorescent aptasensor with hemin/G-quadruplex DNzyme for sensitive detection of lysozyme using rolling circle amplification and strand hybridization. *Biosens. Bioelectron.* **2017**, *87*, 18–24. [CrossRef] [PubMed]
30. Schrödinger, E. Über das Verhältnis der Vierfarben zur Dreifarbentheorie. *Sitz. Akad. Wiss. Math. Kl.* **1925**, *134*, 471–490.
31. Kuehni, R.G. *Color Space and Its Divisions*; John Wiley and Sons: Hoboken, NJ, USA, 2003; ISBN 3175723993.
32. Mahy, M.; Van Eycken, L.; Oosterlinck, A. Evaluation of Uniform Color Spaces Developed after the Adoption of CIELAB and CIELUV. *Color Res. Appl.* **1994**, *19*, 105–121. [CrossRef]
33. Wojciech, M.; Maciej, T. Color difference Delta E—A survey. *Mach. Graph. Vis.* **2011**, *20*, 383–411.
34. Magubane, S.E.; Mlambo, M.; Mabaso, M.H.; Muthwa, S.F.; Kruger, H.G.; Mdluli, P.S. Optimization of CIEL*a*b*/Yxy colour system for colorimetric devices fabricated with gold nanoparticles. *J. Mol. Struct.* **2019**, *1191*, 271–277. [CrossRef]
35. Mbambo, A.T.; Kruger, H.G.; Mdluli, P.S.; Madikizela, L.M. Fabrication and application of a gold nanoparticle-based colorimetric device for the determination of NaCl in seawater and estuarine water. *J. Nanoparticle Res.* **2019**, *21*, 135. [CrossRef]
36. Chaisiwamongkhol, K.; Labaidae, S.; Pon-in, S.; Pinsrithong, S.; Bunchuay, T.; Phonchai, A. Smartphone-based colorimetric detection using gold nanoparticles of sibutramine in suspected food supplement products. *Microchem. J.* **2020**, *158*, 105273. [CrossRef]
37. Lee, P.C.; Meisel, D. Adsorption and surface-enhanced Raman of dyes on silver and gold sols. *J. Phys.Chem.* **1982**, *86*, 3391–3395. [CrossRef]
38. OIV. *Compendium of International Methods of Wine and Must Analysis*; OIV: Paris, France, 2014.
39. Hunt, R.W.G.; Pointer, M.R. *Measuring Colour*; John Wiley and Sons: Chichester, UK, 2011.
40. McDonald, R.; Smith, K.J. CIE94—A new colour-difference formula. *J. Soc. Dye. Colour.* **1995**, *111*, 376–379. [CrossRef]
41. ISO. *INTERNATIONAL STANDARD, Graphic Technology—Process Control for the Production of Half-Tone Colour Separations, Proof and Production Prints*; ISO 12647-2; ISO: Geneva, Switzerland, 2004.
42. Wiernik, P.H.; Serpick, A.A. Clinical Significance of Serum and Urinary Muraminidase Activity in Leukemia and Other Hematologic Malignancies. *Am. J. Med.* **1969**, *46*, 330–343. [CrossRef]
43. Farr, M.; Wardle, E.N.; Wilkinson, K. Lysozymuria in diabetes. *Br. Med. J.* **1976**, *1*, 624–625. [CrossRef]
44. Edwards, J.J.; Tollaksen, S.L.; Anderson, N.G. Proteins of human urine. III. Identification and two-dimensional electrophoretic map positions of some major urinary proteins. *Clin. Chem.* **1982**, *28*, 941–948. [CrossRef]
45. De Paoli Lacerda, S.H.; Park, J.J.; Meuse, C.; Pristinski, D.; Becker, M.L.; Karim, A.; Douglas, J.F. Interaction of gold nanoparticles with common human blood proteins. *ACS Nano* **2010**, *4*, 365–379. [CrossRef] [PubMed]
46. Sen, T.; Mandal, S.; Haldar, S.; Chattopadhyay, K.; Patra, A. Interaction of gold nanoparticle with human serum albumin (HSA) protein using surface energy transfer. *J. Phys. Chem. C* **2011**, *115*, 24037–24044. [CrossRef]
47. Berggard, I. The plasma proteins in normal urine. *Nature* **1960**, *187*, 776–777. [CrossRef] [PubMed]
48. Shrivastav, T.G.; Kariya, K.P.; Prasad, P.K.V.; Chaube, S.K.; Kumar, D. Development of enzyme-linked immunosorbent assay for estimation of urinary albumin. *J. Immunoass. Immunochem.* **2014**, *35*, 300–313. [CrossRef] [PubMed]

nanomaterials

Article

Lysozyme–AuNPs Interactions: Determination of Binding Free Energy

Axel Gomes [1], Jose M. Carnerero [1], Aila Jimenez-Ruiz [1], Elia Grueso [1], Rosa M. Giráldez-Pérez [2] and Rafael Prado-Gotor [1,*]

[1] Department of Physical Chemistry, Faculty of Chemistry, University of Seville, 41012 Seville, Spain; axelgomes314@hotmail.fr (A.G.); jcarnerero2@us.es (J.M.C.); ajimenez28@us.es (A.J.-R.); elia@us.es (E.G.)
[2] Department of Cellular Biology, Physiology and Immunology, Faculty of Science, University of Córdoba, 14014 Cordoba, Spain; rgiraldez@uco.es
* Correspondence: pradogotor@us.es

Abstract: Investigation and optimization of lysozyme (Lys) adsorption onto gold nanoparticles, AuNPs, were carried out. The purpose of this study is to determine the magnitude of the AuNPs–lysozyme interaction in aqueous media by simple spectrophotometric means, and to obtain the free energy of binding of the system for the first time. In order to explore the possibilities of gold nanoparticles for sensing lysozyme in aqueous media, the stability of the samples and the influence of the gold and nanoparticle concentrations in the detection limit were studied. ζ potential measurements and the shift of the surface plasmon band showed a state of saturation with an average number of 55 Lys per gold nanoparticle. Lysozyme–AuNPs interactions induce aggregation of citrate-stabilized AuNPs at low concentrations by neutering the negative charges of citrate anions; from those aggregation data, the magnitude of the interactions has been measured by using Benesi–Hildebrand plots. However, at higher protein concentrations aggregation has been found to decrease. Although the nanocluster morphology remains unchanged in the presence of Lys, slight conformational changes of the protein occur. The influence of the size of the nanoclusters was also investigated for 5, 10, and 20 nm AuNPs, and 10 nm AuNPs was found the most appropriate.

Keywords: binding free energy; gold nanoparticles; lysozyme; nanoparticle size; surface plasmon resonance

Citation: Gomes, A.; Carnerero, J.M.; Jimenez-Ruiz, A.; Grueso, E.; Giráldez-Pérez, R.M.; Prado-Gotor, R. Lysozyme–AuNPs Interactions: Determination of Binding Free Energy. *Nanomaterials* **2021**, *11*, 2139. https://doi.org/10.3390/nano11082139

Academic Editor: Yurii K. Gun'ko

Received: 29 July 2021
Accepted: 19 August 2021
Published: 22 August 2021

1. Introduction

Nanoparticles are characterized by the properties of the metal cluster core, but also by the organic molecules that constitute the monolayer, that is, the capping agents. Gold colloids (AuNPs) are the most stable metal nanoparticles with promising applications because of their electrical, optical, magnetic, and catalytic properties [1]. Besides, gold nanoparticles can be functionalized with a wide variety of structural units by using simple chemical transformations [2–5]. From the first preparation of gold colloids by Faraday in 1857, research on the use of gold nanoparticles has been an area of broad interest, a fact which is reflected by an exponential growth in the number of publications in the recent years. They are considered largely nontoxic, even though this non-toxicity is strongly dependent on their size and recent publications have reported conflicting data [6]. They are also stable, conductive, catalytically active, and electron dense [7]. Among their many qualities, one stands out greatly: AuNPs possess unique optical properties in terms of extremely high extinction coefficients, distance–dependent color, and outstanding fluorescence quenching ability, popularizing them for biosensor development [8]. Owing to those inherent optical properties (AuNPs possess extinction coefficients which are over 1000 times larger than those of organic dyes) certain chemical interactions lead to aggregation-induced color changes allowing for visual detection and quantification of those interactions [9]. Works dealing with protein detection using AuNPs aggregation-based assays, where both label free and functionalized approaches have been explored, are constantly emerging. The simplicity of these methods should not be ignored; one of their strongest points is the low cost that stems from not requiring sophisticated equipment. In aqueous solution AuNPs can appear with different colors (red, blue, purple) and have a broad absorption peak in the range 500–550 nm, depending on the size of the cluster. This peak is called the surface plasmon resonance (SPR) band [10]. This band depends on several factors and therefore making an analysis is not easy. However, for particles smaller than 30 nm it can be simplified since the phenomenon of resonance when light of adequate wavelength reaches the surface of the nanomaterial produces dipoles and not multipoles, as happens with larger ones. The shape, size, and chemical nature of the nanoparticles are factors that affect the plasmon band, as well as the physical–chemical conditions of the medium in which they are found [11]. This fact makes it possible to analyze the interaction of AuNPs with a ligand in solution if aggregation of the clusters takes place.

Among studies using gold nanoparticles as biosensors, those directed toward the detection of proteins are particularly important. The number of possible proteins that in aqueous media can and have shown affinity for gold nanoparticles, and that are able to induce aggregation and the aforementioned color change is vast and fast-growing. Among them, lysozyme (lys) has proven to be a specially interesting case given its medical relevance: it is virtually absent in the finally excreted urine of healthy subjects, but a urinary excretion of elevated levels of lysozyme (lysozymuria) suggests the existence of tubular disfunction [12]. Increased Lys concentration in urine and serum is associated with leukemia [13], renal diseases [14], and meningitis [15]. Although the present study does not focus on biological systems, instead being limited to lysozyme solutions in water, we believe that a good understanding of the phenomena being reported is essential in order to allow for a future extrapolation of the system to more complex cases, where interferents such as ions or other proteins can be present, as are human body fluids. Only by having a good knowledge of the effects that are directly and exclusively caused by lysozyme on gold nanoparticle systems can the influence of the aforementioned interferents be accounted for and compensated, and to our knowledge no such studies have been carried out as of today. Therefore, we have developed a rigorous study of the interaction of lysozyme with the surface of gold nanoparticles, paying special attention to those factors that can directly impact the performance of the gold colloid systems in potential lysozyme detection systems, such as the stability of the samples, the influence of the nanoparticle size, and the correct way to express gold concentration in the system in order to achieve the most precise and reproducible results. Besides, as a result of our studies, the free energy of binding for the 10 nm AuNPs/Lysozyme interaction in water was determined by surface plasmon resonance (SPR) and a deconvolution method, a procedure that to our knowledge has never been described.

2. Materials and Methods

2.1. Materials

All chemicals were of Anal. R. Grade. Gold colloid 5 nm (G1402), gold colloid 10 nm (G1527), gold colloid 20 nm (G1652), and lysozyme (L6876) were purchased from Sigma and NaCl from Merck. Solutions were prepared with deionized water, its conductivity being less than 10^{-6} Sm^{-1}.

2.2. UV-Vis Spectra

The spectra of the AuNPs in the presence and in the absence of lysozyme (Lys) were recorded with a Cary 500 spectrophotometer at 298.2 K from 400 to 700 nm. Nanoparticle concentrations were: [AuNPs] = 8.22×10^{-10} M or [AuNPs] = 3.28×10^{-9} M (exceptions are indicated in the figure legends). Lysozyme concentrations ranged from 10^{-8} M to 10^{-3} M. Total of 200 μL (or 800 μL, see Results and Discussion) of AuNPs 8.22×10^{-9} M was added into a clean quartz cuvette (optical path 10 mm). A timer was started simultaneously as 1800 μL (or 1200 μL) of protein solution of variable concentrations was mixed with the nanosystem solution in the cuvette, which was then reversed several times and then inserted into the cuvette holder in the UV-vis spectrometer. Photograph corresponding to Figure S6 was taken with a Canon 5-50 digital camera.

2.3. Zeta Potential Experiments

Zeta potential measurements were carried out at 298.2 K with a Zetasizer Nano ZS Malvern Instruments Ltd (UK). A DTS 1060 polycarbonate capillary cell was used. Figure S1 depicts zeta potential distributions of the 10 nm gold colloids with a negative charge of about -33.1 ± 1.6 mV in water, which is sufficient to keep the particles from interacting with each other and therefore prevent aggregation of the sample, maintaining a stable particle size. Measurements were performed in Lys concentrations range from 10^{-8} M to 3×10^{-4} M. [AuNPs] = 8.22×10^{-10} M in each case.

2.4. Circular Dichroism (CD) Spectra

Electronic CD spectra were recorded in a BioLogic Mos-450 spectropolarimeter. A standard quartz cell of 10 mm path length was used. The spectra were expressed in terms of ellipticity. Scans were taken from 190 nm to 250 nm and for each spectrum 5–10 runs were averaged at a constant temperature of 298.2 K with a 5 min equilibration before each scan. [AuNPs] = 8.22×10^{-10} M. [Lys] = 10^{-8} M.

Deconvolution of CD spectra was done by using CDNN software (version 2) with the standard basis set of 33 spectra [16].

2.5. Kinetic Measurements

Kinetic runs were carried out in a stopped-flow spectrophotometer from Applied Photophysics. The reaction was monitored by following the changes in absorbance of the surface plasmon band, SPR, at 565 nm, in the presence of [Lys] = 4×10^{-7} M and [AuNPs] = 3.28×10^{-9} M. A solution of the same AuNPs concentration in water was used as reference. The temperature was maintained at 298.2 ± 0.1 K.

2.6. TEM Measurements

For TEM examinations, a single drop (10 μL) of the aqueous solution of the gold nanoparticles was placed on a carbon film-coated copper grid which was then left to dry in air for several hours at room temperature. TEM analysis was carried out in a Philips CM 200 electron microscope working at 200 kV. [AuNPs] = 8.22×10^{-10} M; [NaCl] = 0.1 M; [Lys] = 4×10^{-8} M.

2.7. Deconvolution

Deconvolution of experimental spectra was carried out by fitting to Voigt functions, which best reproduced the SPR of free nanoparticles. Fityk software (0.9 version) was used to carry out the fit [17]. Spectra were fitted to a three-band model, where the free AuNPs band was always found at around 521 nm; the linked AuNPs band appeared at 561 nm for almost all cases except those where the aggregation degree was too small to be significant. A third broad band was fixed at 300 nm in order to account for both residual Au^{3+} absorbance and for the light scattering phenomena linked to the presence of nanoparticles in solution [18].

3. Results and Discussion

3.1. Stability and Detection Limit Considering the Red Shift of the SPR

Figure 1 shows the surface plasmon band for two different batches of 5 nm colloidal gold (ref. G1402). The spectra correspond to a mixture of 200 μL of commercial gold with 1800 μL of distilled water. The black solid line corresponds to the batch Lot. SLBB2962V with an average size of 4.8 nm and a gold concentration of 67 μg/mL. The red dashed line corresponds to the batch Lot. 061M6031 with an average size of 5.1 nm and a gold concentration of 45 μg/mL. In order to highlight the influence of gold concentration and nanoparticle size over the SPR band, gold concentrations have not been corrected. It can be clearly seen that a small difference in size (0.3 nm) promotes a change in the maximum of the SPR, ($\Delta\lambda$ = 9 nm), and in the optical density (O.D): Δ(O.D) ~ 18%.

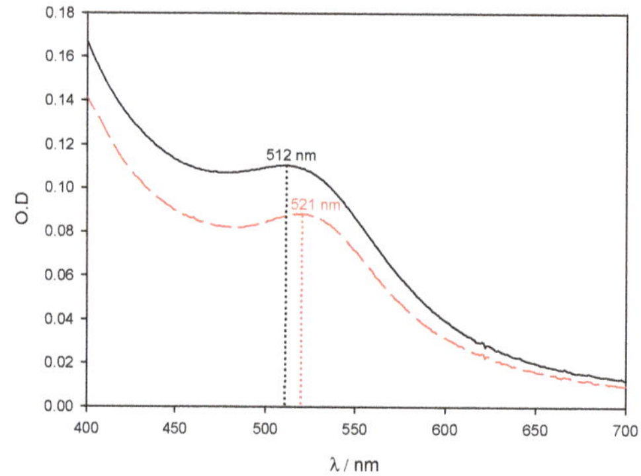

Figure 1. Experimental spectra of AuNPs solutions on the visible region. Commercial gold nanoparticles were used as received. Black solid line corresponds to 4.8 nm AuNPs and gold concentration of 67 μg/mL. Red dashed line corresponds to 5.1 nm AuNPs and gold concentration of 45 μg/mL.

Figure 2 shows the SPR stability corresponding to 10 nm AuNPs in the presence of lysozyme 3 × 10^{-4} M. When compared to the SPR of lone AuNPs, λ_{max} shifts to slightly higher values over time when the protein is present in solution; those changes are accompanied by a slight darkening of the red tint of the nanoparticles. Although during the first minutes λ_{max} changes some nanometers, as the optical density does, after the first 10 min. and at least during 1 hour the system's changes take place at a very slow speed, which allows for the 10 minute mark to be taken as a reference in all measurements. For this reason, in further experiments, the values of λ_{max} and O.D. where obtained 10 min after the addition of the protein to the colloidal gold. In order to control the stability of the aggregates, measurements were made at multiple fixed [AuNPs]/[Lys] relationships. Figure 3 reflects compliance with the Lambert–Beer law in the range [AuNPs]/[Lys] = 4.10 × 10^{-9} M/2.50 × 10^{-7} M – 8.22 × 10^{-10} M/5.00 × 10^{-8} M. This implies that while we worked at low concentrations of gold nanoparticles so as to determine the potential lowest detection limits of a lysozyme-sensing system in water, the optical density of the solutions also provides valuable information to this end, and could be explored as an alternate mean for lysozyme quantification.

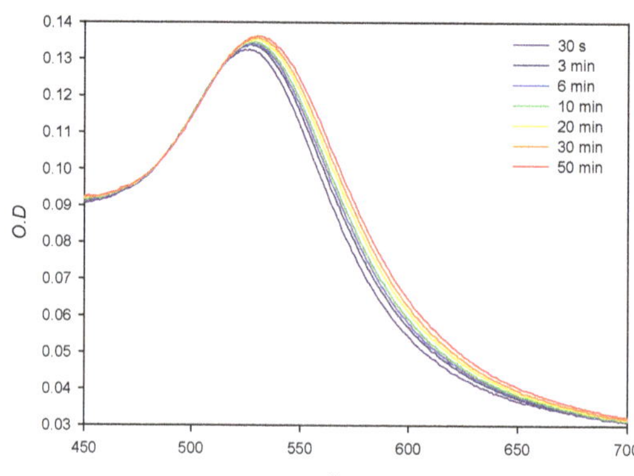

Figure 2. Experimental spectra of 10 nm AuNPs in the presence of Lys 3 × 10^{-4} M. [AuNPs] = 8.22 × 10^{-10} M.

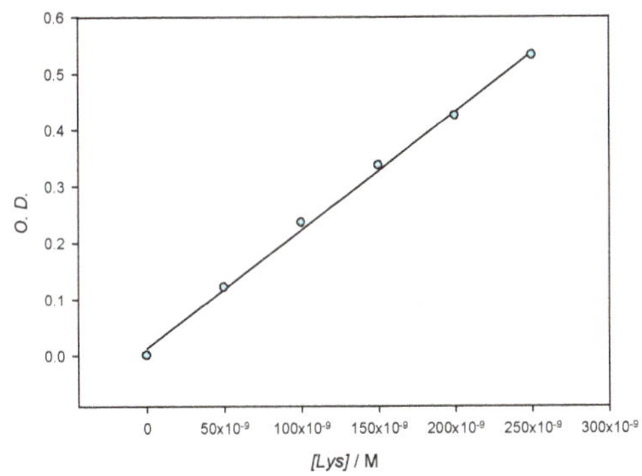

Figure 3. Optical density at different Lys concentrations for a constant ratio [AuNPs]/[Lys] = 0.0164 measured at 565 nm.

The influence of the protein concentration in experimental absorbance λ_{max} measurements is shown in Figure 4A. As can be seen, at low enough concentrations of lysozyme, the SPR experiences a red shift that peaks at [Lys] = 4×10^{-8} M; this effect is reflected in Table 1, which shows the values of $\Delta\lambda$ at different Lys concentrations when keeping the gold concentration constant. Aggregation phenomena are more marked at low concentrations of the protein, yielding a detection limit in the nanomolar range, around 1.3×10^{-9} M. It is also possible to employ the O.D. and consider the degree of aggregation of the nanoclusters in function of the protein concentration (A_{569}/A_{518}; where A_λ stands for absorbance values at λ nm). However, the loss of accuracy can be seen in Figure 5B when compared with Figure 5A. In this study, we have worked with very low gold concentrations, even at the risk of having O.D. $\approx$ 0.1 in order to achieve high accuracy in the value of λ_{max}. Our main goal is to explore the possibilities of using the pronounced λ_{max} variations as an alternative signal instead of A_{569}/A_{518} and thus improve the perception of the system changes taking place during the formation of the AuNPs/Lysozyme complex.

The detection limit for lysozyme concentrations in water has been obtained from 3S/D, where S is the standard deviation of the measurements ($\pm$0.37) and D the slope of the calibration line (see Figure 5A). Meanwhile, the lower limit of quantification (10 S/D) is 4.3×10^{-9} M. Those detection limits, which have been achieved with non-functionalized 10 nm AuNPs, are remarkable when compared to previous studies, even those that use functionalized nanoparticles [19].

Table 1. Values of $\Delta\lambda$/nm at different protein concentrations. [AuNPs(10 nm)] = 8.22×10^{-10} M.

[Lys]/M	$\Delta\lambda$/nm	[Lys]/M	$\Delta\lambda$/nm
1.0×10^{-8}	4	5.0×10^{-7}	24
1.5×10^{-8}	9	1.0×10^{-6}	22
2.0×10^{-8}	15	1.0×10^{-5}	21
2.5×10^{-8}	16	5.0×10^{-5}	15
3.0×10^{-8}	23	1.0×10^{-4}	13
4.0×10^{-8}	28	3.0×10^{-4}	11
5.0×10^{-8}	27	1.0×10^{-3}	8
1.0×10^{-7}	23	-	-

3.2. *Saturation of the System AuNPs/Lys: Structural Characterization*

If data of λ_{max} in Table 1 are carefully analyzed, it can be observed that aggregation reaches a state of saturation. Indeed, at higher concentrations of protein, aggregate size begins to decrease; this fact becomes apparent when analyzing the absorbance spectra in Figure 4B, which shows a λ_{max} shift to lower values for lysozyme concentrations higher than 4×10^{-8} M. At a given point the nanoparticle surface is saturated and protected all around by a layer of Lys, which prevents aggregation due to steric hindrance and repulsion forces. It is important to note that the concentration of Lys from which λ_{max} starts to decrease matches the absorbance ratio (see Figure S2 in Supporting Information). Considering the AuNPs concentration and the concentration of Lys at which saturation is observed, the average number of protein molecules per nanoparticle has been determined to be 55. Lysozyme has been found to have an estimated size of 3.0 nm × 3.0 nm × 4.5 nm [20]. If the ellipse formed by the longer and one of the shorter diameters, which would have an area of 10.6 nm^2, is assumed to fully interact with the 10 nm AuNPs surface (314.2 nm^2), then around 30 lysozyme molecules would fit. If, instead, the shorter diameters are considered, then each lysozyme would take up a circle of 7.07 nm^2, which would allow around 44 lysozyme molecules to fit. Those theoretical approximations are in accordance with the reported experimental data, since the lysozyme molecules do not need to fully use one of their sides in order to interact with the AuNPs.

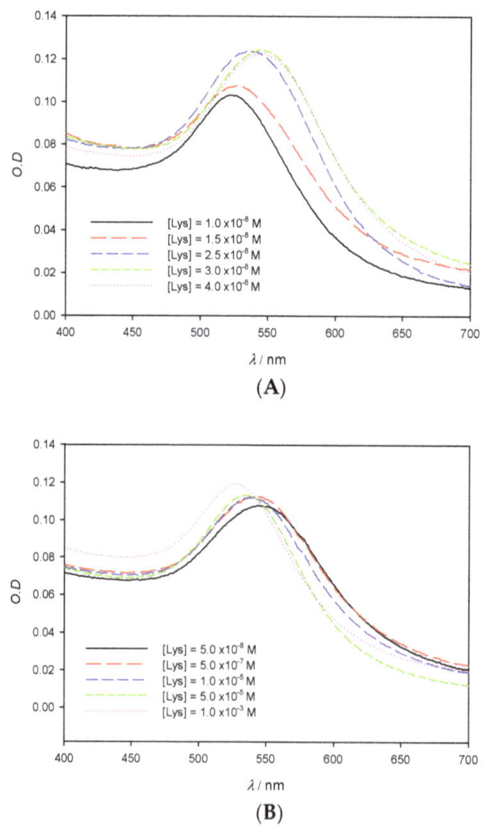

Figure 4. Experimental spectra of 10 nm AuNPs solutions on the visible region showing the red (**A**) and blue (**B**) shift in function of the Lys concentration. [AuNPs] = 8.22 × 10^{-10} M.

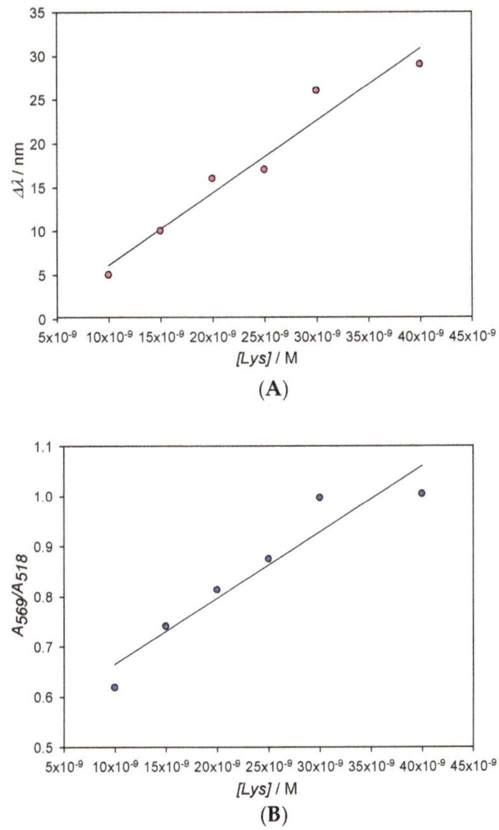

Figure 5. Changes of the maximum of the SPR (**A**) and the A_{569}/A_{518} ratio (**B**) with Lys concentration. From Figure 5A, a value of 1.3×10^{-9} M for the detection limit (L.D) has been obtained, with L.D. = 3S/D being D the slope of the line and S the standard deviation of the measurements.

Figure 6 shows three arrays of TEM image photographs that help confirm that the aggregation process takes place as previously stated. It shows images corresponding to solutions in the presence of 10 nm AuNPs (λ_{max} = 518 nm) (A), in the presence of NaCl 0.1 M as the reference aggregation process (λ_{max} = 616 nm) (B), and in the presence of Lys concentration at which the observed shift of λ_{max} is maximum (λ_{max} = 546 nm, see Table 1, [Lys] = 4×10^{-8} M) (C). Figure 6C,D clearly shows the aggregation of nanoparticles in the presence of Lys without any apparent changes in nanoclusters morphology.

In order to further explore the nature of AuNPs–Lys interactions, polarization spectroscopy measurements were carried out (Figure 7). Despite the clear shape difference observed between the lone lysozyme and the lysozyme–AuNPs CD spectra, the deconvolution with CDNN software did not show appreciable changes in the percentages of α-helix, β-turn, and random coil structures present in lone lysozyme in relation to those that form in the presence of AuNPs (see Table S1). Those results are not unexpected due to lysozyme presenting four internal disulphide bonds that confer it a certain degree of stiffness and greatly stabilize its secondary structure, even in the presence of a ligand such as AuNPs. Since there is no appreciable change in the secondary structure of the protein, which strongly depends on those disulphide bonds as well as on intra-molecular hydrogen bonds, we can conclude that the interaction with the AuNPs does not break nor alter either of those links. It is also to note that peptide bonds in proteins are well-known to cause two strong CD bands in the far-UV: one corresponding to $\pi \rightarrow \pi^*$ electronic transitions, located around 190 nm, and the other to n → π*, which appears at around 210 nm [21]. Those transitions are clearly visible in the lone lysozyme CD spectra (red line), and show a shift to higher wavelengths, implying lower transition energies, when AuNPs are added (blue line) with a fine structure appearing around 195 nm. We can conclude that even though sulfur atoms have been known to bond strongly with gold [22] the added stability of the disulphide bonds causes AuNPs to preferably interact with the peptidic bond atoms, thus not affecting the secondary structure of the protein. This effect is in agreement with previous reports; for example, 2-(10-mercaptodecyl) malonic acid functionalized 2 nm core gold nanoparticles (anionic AuDA), which form a high affinity complex with Lys, have been found to actually promote re-folding of denatured proteins by shielding them from inter-protein hydrophobic forces; denatured proteins that were treated with AuDA have been shown to present conformations that mimicked those of the native proteins, further proving the point that nanoparticles do not cause significant alterations of the secondary structure of the protein [23]. In any case, the CD results presented in the present study should be considered as an indication, not definitive proof in relation with the secondary structure of the protein.

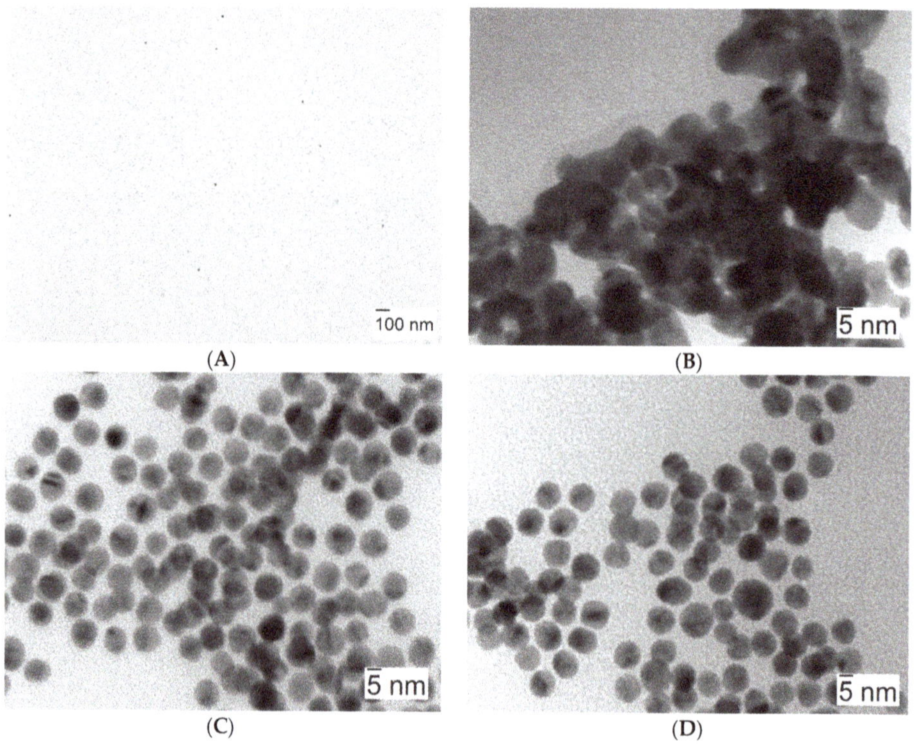

Figure 6. TEM images corresponding to 10 nm AuNPs. (**A**) [AuNPs] = 8.22×10^{-10} M, in water (shown at a lesser magnification in order to better convey nanoparticle dispersion). (**B**) [AuNPs] = 8.22×10^{-10} M in the presence of [NaCl] = 0.1 M. (**C**,**D**) [AuNPs] = 8.22×10^{-10} M in the presence of [Lys] = 4×10^{-8} M.

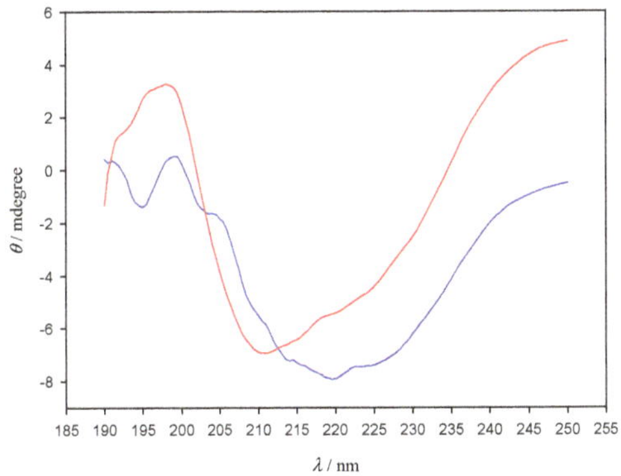

Figure 7. CD spectra of Lys in the absence (red line) and in the presence of 10 nm gold nanoparticles (blue line). [AuNPs] = 8.22×10^{-10} M. [Lys] = 10^{-6} M.

The stability of different kinds of nanoparticles strongly depends on the valence of the counterions in the solution due to the electrokinetic or ζ potential, this is, the difference between the compact layer potential and the diffuse potential. In this sense, the magnitude of the measured ζ potential is an indication of the repulsive force that prevents nanoparticle aggregation and actually allows the colloidal gold solutions to be stable, and as such can be used to predict the long-term stability of the nanoparticle suspension.

Table 2 shows ζ potential values at different Lys concentrations when measured at a constant gold nanoparticles concentration. In the absence of the protein a value of −33 mV (see S1 in supporting information) is sufficient to keep the AuNPs away from each other, thus achieving a stable particle size. ζ potential corresponding to the Lys in the absence of AuNPs indicates that the protein is positively charged at neutral pH (+8.71 mV). As can be seen in Table 2, as the protein concentration increases, the zeta potential becomes less negative, as is to be expected in response to the association of Lys molecules on the gold surface. The partial neutralization of the AuNPs surface decreases the repulsion forces between nanoparticles, facilitating the aggregation. At a concentration around 10^{-5} M a charge reversal occurs, and ζ potential becomes positive. Now AuNPs/Lys entities have a positive average charge and actually begin to repel each other. The situation of maximum aggregation observed from the shifts of the SPR appears at a protein concentration ([Lys] = 4×10^{-8}–5×10^{-8} M) which agrees well with the Lys concentration at which the ζ potential approaches zero. These two behaviors can be described by Scheme 1.

Table 2. Values of ζ potential/mV at different protein concentrations. [AuNPs (10 nm)] = 8.22×10^{-10} M. ζ Potential = +8.71 mV for Lys in the absence of AuNPs.

[Lys]/M	ζ Potential/mV
0	−33.0
1.0×10^{-8}	−30.1
2.0×10^{-8}	−29.9
3.0×10^{-8}	−22.6
4.0×10^{-8}	−13.8
5.0×10^{-8}	−10.5
1.0×10^{-6}	9.4
5.0×10^{-5}	15.4
3.0×10^{-4}	10.1

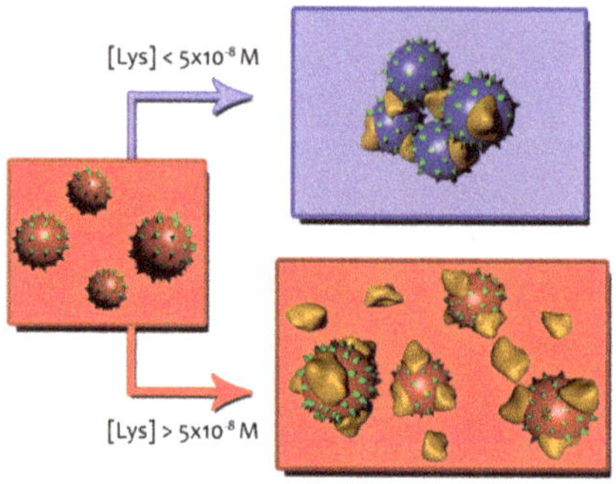

Scheme 1. Schematic representation of AuNPs/Lys entities.

In order to further understand the system, information about the kinetics and the binding free energy of the AuNPs–Lys interaction has been obtained. Figure 8 shows optical density changes versus time at λ = 565 nm for [AuNPs] = 3.28 $\times$ 10^{-9} M and [Lys] = 10^{-7} M. Although, for stability reasons, the maxima of the SPR were collected 10 min after mixing the colloidal gold with the protein, it can be seen that, in seconds, it reaches 90% aggregation being only 0.5 seconds necessary to reach an aggregation of 50%. Experimental data cannot be fitted to a monoexponential equation, which evidences that the mechanism of AuNPs–Lys interaction is a complex one.

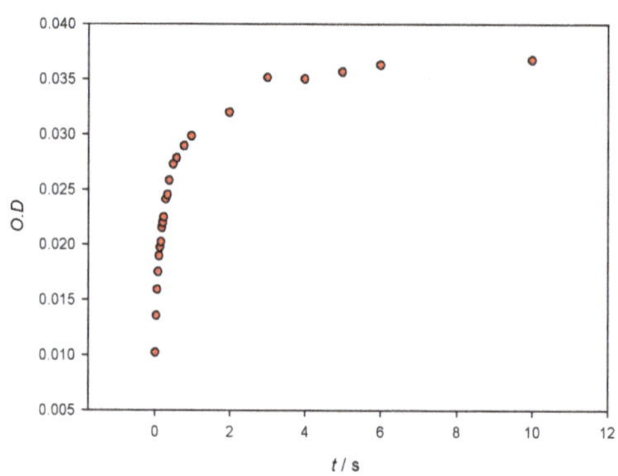

Figure 8. Optical density changes at λ = 565 nm versus time when mixing 10 nm AuNPs with Lys. Concentrations in the stopped flow cuvette are [AuNPs] = 3.28 $\times$ 10^{-9} M and [Lys] = 10^{-7} M.

3.3. *Influence of the Nanoparticle Size and Gold Concentration. Determination of Binding Free Energy of Lysozyme to Gold Nanoparticles*

A common misconception in gold nanoparticle research papers is that concentration of the colloids is quite frequently expressed in terms of starting gold salt concentrations, so the influence of the particle size is not taken into account. While this approach is correct when working with colloids of a given, monodisperse particle size, in order to study the influence of particle radius into the AuNPs–Lys system working with particle concentrations instead of gold concentrations is a must. As reflected in Scheme 2, varying the particle size at constant gold concentration causes an obvious change in the number of nanosystems in solution. For our purpose of studying the influence of size, it is important to know the number of moles per liter of each nanoparticle size; to this end, the concentration of nanoparticles in each commercial sample was determined as described in the Supporting Information. Our results were practically identical to those of the manufacturer.

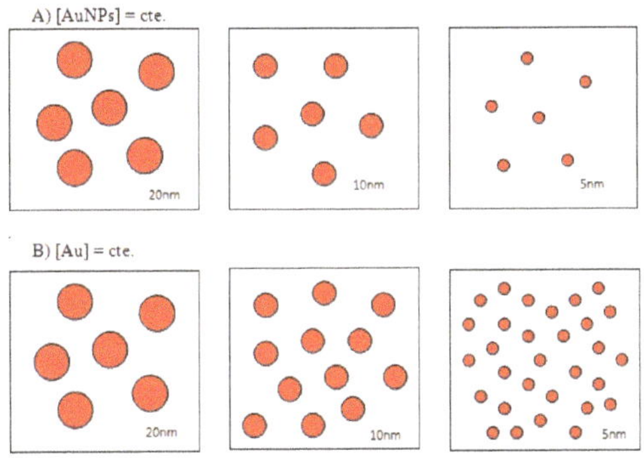

Scheme 2. Schematic representation of 5, 10, and 20 nm AuNPs at constant nanoparticles concentration (**A**) and at constant gold concentration (**B**).

Therefore, to analyze the effect of the nanoparticle size in the colorimetric detection process the number of nanoclusters in solution must be kept constant. This situation is reflected in Figure S3A where the SPR of 5, 10, and 20 nm are represented at identical nanoclusters concentrations and in Figure S3B where the SPR corresponds to solutions of identical gold concentrations. As it can be seen, in order to obtain reproducible optical density data, it is essential to increase the concentration of gold well above those used in the experiments described so far. The influence of colloidal gold based on the nanoparticles' size is a key point in this approach; usually, smaller sized ones are more stable than the larger ones as the latter have higher intensity Van der Waals interactions between them, and these interactions are attractive. Thus, in general, the smaller nanoparticles take longer to aggregate and, once aggregated, may be stable for hours.

The system is stable when working with 20 nm AuNPs and the most dilute solution of Lys (10^{-8} M). However, the observed $\Delta\lambda$ was only 1 nm, instead of the 4 nm shift previously reported when 10 nm AuNPs were employed (see Table 1). On the other hand, at high protein concentrations, the system, unlike 10 nm AuNPs, is no longer stable (see Figure S4) showing a precipitation marked by the continuous red shift of λ_{max} and a sudden drop in absorbance.

As the nanoparticle surface increases, it is necessary to achieve higher protein concentrations in order to aggregate the AuNPs and promote a red shift of the SPR. Thus, a better detection limit is achieved when working with 10 nm AuNPs instead of 20 nm AuNPs. Due to the higher stability of smaller nanoparticles over larger ones, as explained previously, 5 nm AuNPs would be expected to be the most appropriate, but then the ratio Lys/AuNPs should be raised to achieve a reasonable absorbance. The reason is that to compare results at identical concentrations of nanoparticles, 24 µL of 5 nm AuNPs stock solution must be added to 1800 µL of distilled water, compared with 200 µL of 10 nm AuNPs stock solution, and this means having a very low absorbance (see Figure S3A). Figure 9 shows that, instead of 20 nm AuNPs, when working with 5 nm AuNPs and multiplying, for example, the ratio AuNPs/Lys by a factor of 4 relative to 10 nm AuNPs, in order to maintain a constant AuNPs/Lys concentration ratio, the system is stable in the presence of concentrated ([Lys] = 4×10^{-4} M) and dilute solutions of protein (data not shown). Even more, with 5 nm AuNPs, a higher detection limit is achieved ($\Delta\lambda = 7$ nm for the lowest concentration of Lys, [Lys] = 10^{-8} M) maintaining the same value of $\Delta\lambda$ for [Lys] = 10^{-4} M as that obtained with 10 nm AuNPs in the presence of [Lys] = 3×10^{-4} M, that is, $\Delta\lambda = 11$ nm (see Table 1). However, although considering these points 5 nm AuNPs would be the most appropriate to sense the presence of Lys in solution, their wide SPR compared with the better defined SPR of 10 nm AuNPs (see Figure S3B) makes the latter the most appropriate. Results corresponding to 5, 10, and 20 nm are combined into Figure S5.

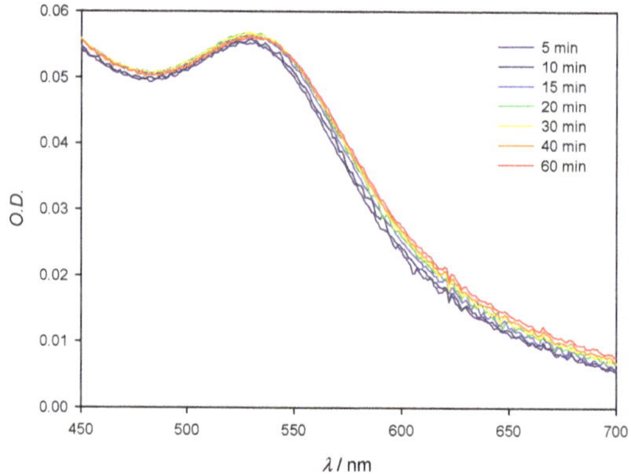

Figure 9. Experimental spectra of 5 nm AuNPs in the presence of Lys. [AuNPs] = 4 × 8.22 × 10^{-10} M; [Lys] = 4 × 10^{-4} M.

With the intention of analyzing the influence of the gold concentration in the AuNPs/Lys system, 10 nm AuNPs have been used at a concentration four times greater than that employed throughout this study. Figure 10 and Table 3 show the obtained results. Note that the red and the blue shift are once again observed, depending on the Lys/AuNPs ratio. As the detection limit worsens (Δλ = 1 nm for [Lys] = 10^{-8} M), larger values of Δλ are achieved at high Lys concentrations, as in this case the saturation of the AuNPs surface is reached at higher protein concentrations. Naturally, as shown in a comparative photograph, the color changes are more intense and defined than when samples of more dilute gold are used (see Figure S6).

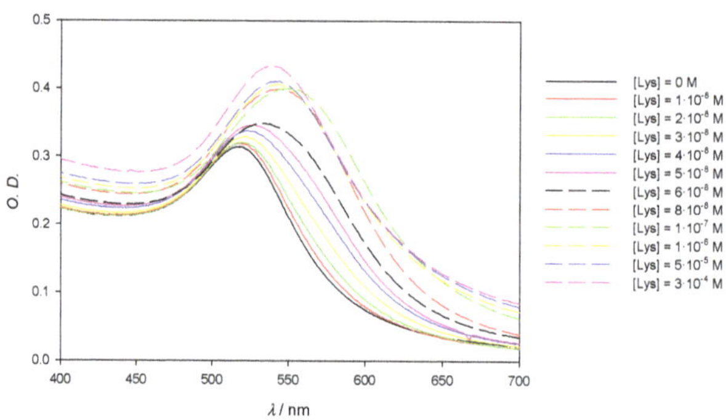

Figure 10. Experimental spectra of 10 nm AuNPs solutions on the visible region showing the red and blue shift in function of the Lys concentration. [AuNPs] = 3.28 × 10^{-9} M. The blue arrow points the shift of the SPB according the Lys concentration increases; the red dashed arrow highlights the reversal of the band at highest concentrations.

Table 3. Values of Δ λ/nm at different protein concentrations. [AuNPs (10 nm)] = 3.28 × 10^{-9} M.

[Lys]/M	Δλ/nm	[Lys]/M	Δλ/nm
1.0 × 10^{-8}	1	8.0 × 10^{-8}	27
2.0 × 10^{-8}	2	1.0 × 10^{-7}	32
3.0 × 10^{-8}	4	1.0 × 10^{-6}	27
4.0 × 10^{-8}	5	5.0 × 10^{-5}	27
5.0 × 10^{-8}	10	3.0 × 10^{-4}	21
6.0 × 10^{-8}	16	-	-

From the spectra shown in Figure 10, deconvolution procedures were carried out in order to obtain a better view of both SPRs: the one corresponding to non-interacting, free AuNPs, which show a band whose peak is already known to be located at around 521 nm; and the one from aggregated nanoparticles, whose width makes it hard to determine the exact location of its absorbance maxima. The deconvoluted peak sum shows a near perfect agreement with experimental data when the fit is done with a three peak model, as shown in Figure 11; the third peak is fixed at 300 nm and accounts for both residual Au^{3+} salts (which show a broad band centered around that wavelength) and for the dispersion effects that happen in the near-UV and are inherent to gold nanoparticle solutions [18]. It is also important to note that the λ_{max} of the aggregation band may shift with increasing concentrations of lysozyme, as the size distribution of aggregates becomes more disperse and the band widens. However, this effect was not observed in our system, as reflected in Table 4, and the aggregation band maxima remained at 561 ± 1 nm for all measurements except [Lys] = 2×10^{-8} M, where the aggregation degree is still very low.

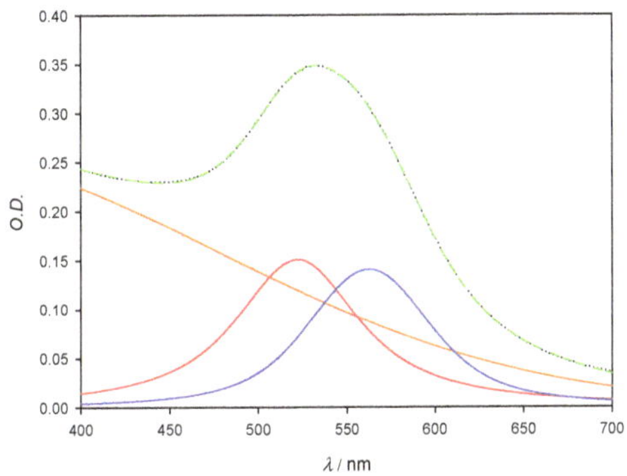

Figure 11. Deconvolution plot for [AuNPs] = 6×10^{-8} M. Orange line corresponds to residual Au^{3+} salts; red line corresponds to the SPR of free gold nanoparticles and peaks at 522 nm; blue line shows the SPR that arises from the formation of the gold aggregates. Black (dotted) line corresponds to the experimental spectra and green (dashed) line corresponds to the sum of all deconvoluted bands.

Table 4. Values of λ_{max}/nm for both free and aggregated AuNPs at different protein concentrations.

[Lys]/M	λ_{max} (Free AuNPs)/nm	λ_{max} (Aggregated AuNPs)/nm
1.0×10^{-8}	521	0
2.0×10^{-8}	521	559
3.0×10^{-8}	521	561
4.0×10^{-8}	521	562
5.0×10^{-8}	521	561
6.0×10^{-8}	521	562
8.0×10^{-8}	521	562
1.0×10^{-7}	521	561

The evolution, both in area and in maxima intensity, of the two deconvoluted peaks is directly linked to the fraction of AuNPs that are bound to lysozyme for each concentration, although the direction in which this change takes place shifts in accordance to the aggregation–disaggregation effects mentioned earlier. When aggregation phenomena are prevalent, the intensity of the aggregated deconvolution band at 561 nm grows with increasing [Lys]. Past the turning point of 1×10^{-7} M, increasing [Lys] causes the aggregation band to shrink, because the excess lysozyme has the effect of preventing aggregation, as previously discussed.

Since the absorbance of the aggregated AuNPs band (and by extent, that of the free AuNPs band) is directly linked to the concentration of aggregated AuNPs, those absorbance data offer a reliable way of determining the binding constant, and by extent the binding free energy, ΔG^0, of the AuNPs/Lys complex by using a simplified reaction scheme:

$$AuNPs + Lys \overset{K_b}{\leftrightarrow} AuNPs/Lys \quad (1)$$

where K_b is the binding constant of our system.

The stable position of the absorbance peak allows for the binding constant to be determined by using the Benesi–Hildebrand equation [24], which uses absorbance measurements at a fixed wavelength to monitor the formation of a colored complex during a chemical reaction:

$$\frac{[AuNPs]_0}{Abs} = \frac{1}{K_b\varepsilon\,[Lys]_0}\frac{1}{\varepsilon} \quad (2)$$

where K_b is the formation constant of the complex, and ε its molar extinction coefficient. For this model, it is assumed that the absorbance band of the complex does not shift during the formation process.

Figure 12 shows a Benesi–Hildebrand plot of the absorbance measured at 561 nm for [Lys] = 4×10^{-8} M to 1×10^{-7} M; lower concentration points showed slight deviations of linearity, probably due to the low concentration of aggregated nanoparticles in solution. From the fit, the value obtained for K_b was 1.64×10^7 M^{-1}, which corresponds to a free energy of binding $\Delta G^0 = -41.2$ kJ mol^{-1}. Extinction coefficient (ε) of the aggregated AuNPs was also calculated, finding a value of 1.95×10^8 $M^{-1}cm^{-1}$. When the extinction coefficient of 10 nm free AuNPs was estimated through theoretical means [25], a value of 1.03×10^8 $M^{-1}cm^{-1}$ was found, remarkably close to that of our experimental measurements.

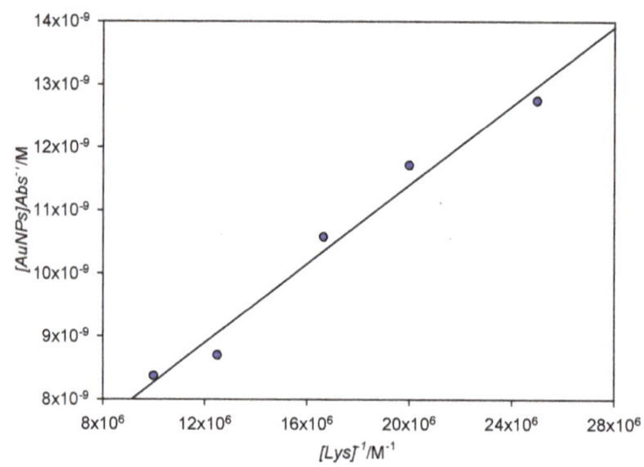

Figure 12. Benesi–Hildebrand plot for the absorbance at 561 nm for [Lys] = 4×10^{-8} $M^{-1} \times 10^{-7}$ M.

In order to corroborate those results, the Benesi–Hildebrand equation was also applied to the area quotient between the aggregated and the free peaks. Since this fitting model was developed in order to be used in cases where the complex formation is directly proportional to the receptor concentration (in our case, [Lys]) by virtue of the Lambert–Beer law, the aggregated/free area ratio would also need to be directly proportional to [Lys] for the model to be valid. It is known, from the previous fit, that the aggregated peak intensity is directly proportional to [Lys]; knowing that the formation of the aggregate is directly linked to the disappearing of the free SPR, and supposing that no band widening takes place and both the aggregated and free peaks remain roughly the same shape, then it follows that the area of both peaks should be directly proportional to one another and by extension to [Lys], albeit the proportionality constant will not correspond to ε since this constant is linked to absorbance and not area. If the previous statements are true, then the Benesi–Hildebrand plot for the aggregated/free area quotient should also be a straight line; experimental results are shown in Figure 13. As can be seen, the fit of the area quotient to the Benesi–Hildebrand equation is good (R^2 = 0.988). The K_b for this fit is 1.20 × 10^7 M^{-1}, corresponding to $\Delta G^0 = -40.4$ kJ mol^{-1} and so in a really good agreement to that obtained by classical Benesi–Hildebrand representations.

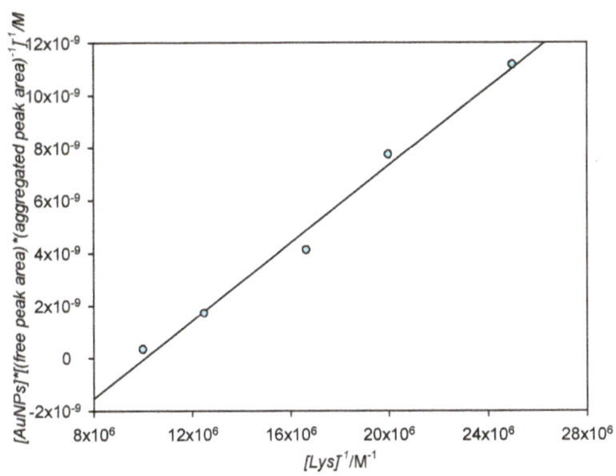

Figure 13. Benesi–Hildebrand plot for aggregated/free area quotients for [Lys] = 4 × 10^{-8} M^{-1} × 10^{-7} M.

4. Conclusions

A thorough study of AuNPs–Lys interaction in aqueous solution has been carried out. Aggregation occurs in seconds, but it has been found that at high protein concentrations the aggregation degree of the AuNPs markedly decreases, a fact confirmed by the blue shift of λ_{max}, the observed color changes of AuNPs/Lysozyme solutions from purple to red, and by the changes of ζ-potential, going from a negative to a positive value as [Lys] increases. The state of saturation has been found to imply an average number of 55 Lys per gold nanoparticle, in good accordance with both the protein and the AuNPs size. Although the nanocluster morphology remains unchanged in the presence of Lys, conformational changes of the protein occur, but the secondary structure of the protein remains unchanged, which indicates that no lysozyme disulphide bonds break from the interaction with AuNPs. From experimental absorbance data, by using Benesi–Hildebrand plots of both A_{561} and deconvoluted free AuNPs/aggregated AuNPs absorbance area quotients, the free energy of binding (ΔG^0) of lysozyme to 10 nm AuNPs has been found to be around −40.8 kJ mol^{-1} with a good agreement between the two fits; to our knowledge, this work is the first instance of deconvolution parameters being used in order to determine the binding constant of lysozyme to gold nanoparticles. Studies of the influence of AuNPs size on the interaction with lysozyme confirm the need of using nanoparticle concentrations instead of gold concentrations when approaching the study of a new nanosystem, and also show that 10 nm AuNPs are the most appropriate for those studies. Working with high gold concentrations does not improve the results achieved, although better defined changes in color are accomplished.

Supplementary Materials: The following are available online at https://www.mdpi.com/article/10.3390/nano11082139/s1, Table S1: Deconvoluted structures of native lysozyme and the AuNPs-lysozyme complex, Figure S1: ζ-Potential of 10 nm AuNPs, Figure S2: (A) Changes of the maximum of the SPB (λ_{max}). (B) Changes of the ratio A_{569}/A_{518} at each protein concentration of Table 1. [AuNPs] = 8.22 × 10^{-10} M, Figure S3: (A) SPB of 5, 10 and 20 nm AuNPs at the same nanoparticle concentration. [AuNPs] = 8.22 × 10^{-10} M. (B) SPB of 5, 10 and 20 nm AuNPs at the same gold concentration. [Au] = 50 μ, Figure S4: Stability study of 20 nm AuNPs in the presence of [Lys] = 10^{-4} M. [AuNPs] = 8.22 × 10^{-10} M, Figure S5: (A) Stability study of 5 nm AuNPs in the presence of [Lys] = 4 × 10^{-4} M. [AuNPs] = 4 × 8.22 × 10^{-10} M; (B) Stability study of 10 nm AuNPs in the presence of [Lys] = 3 × 10^{-4} M. [AuNPs] = 8.22 × 10^{-10} M. (C) Stability study of 20 nm AuNPs in the presence of [LYS] = 10^{-4} M. [AuNPs] = 8.22 × 10^{-10} M, Figure S6: A comparative analysis corresponding to the first nine Lys concentrations of Table 3 in the presence of [AuNPs] = 8.22 × 10^{-10} M and [AuNPs] = 3.28 × 10^{-9} M.

Author Contributions: Methodology, A.J.-R.; validation, E.G.; supervision, R.M.G.-P. and R.P.-G.; validation: E.G; formal analysis, J.M.C. and R.M.G.-P.; investigation, A.G., J.M.C. and A.J.-R.; writing-original draft preparation, R.P.-G.; writing—review and editing, E.G.; funding acquisition, R.P.-G. All authors have read and agreed to the published version of the manuscript.

Funding: This work was financed in part by the Consejería de Educación y Ciencia de la Junta de Andalucía (P08-FQM-03623). We thank European Social Fund and Junta de Andalucía for the postdoctoral fellowship "Contratación de Personal Investigador Doctor (Convocatoria 2019). PAIDI 2020".

Acknowledgments: We thank F. Varela (CITIUS, University of Seville) and J. Quispe for their assistance in obtaining TEM images and ζ potential measurements, respectively.

Conflicts of Interest: The authors declare no conflict of interest.

References

1. Wilcoxon, J. Optical Absorption Properties of Dispersed Gold and Silver Alloy Nanoparticles. *J. Phys. Chem. B* **2009**, *113*, 2647–2656. [CrossRef] [PubMed]
2. Hostetler, M.J.; Green, S.J.; Stokes, J.J.; Murray, R.W. Monolayers in Three Dimensions: Synthesis and Electrochemistry of ω-Functionalized Alkanethiolate-Stabilized Gold Cluster Compounds. *J. Am. Chem. Soc.* **1996**, *118*, 4212–4213. [CrossRef]
3. Ingram, R.S.; Murray, R.W.; Hill, C.; Carolina, N. Electroactive Three-Dimensional Monolayers: Anthraquinone ω-Functionalized Alkanethiolate-Stabilized Gold Clusters. *Langmuir* **1998**, *14*, 4115–4121. [CrossRef]
4. Green, S.J.; Stokes, J.J.; Hostetler, M.J.; Pietron, J.; Murray, R.W. Three-Dimensional Monolayers: Nanometer-Sized Electrodes of Alkanethiolate-Stabilized Gold Cluster Molecules. *J. Phys. Chem. B* **1997**, *101*, 2663–2668. [CrossRef]
5. Templeton, C.; Hostetler, M.J.; Warmoth, E.K.; Chen, S.; Hartshorn, C.M.; Kroshnamurthy, V.M.; Forbes, M.D.E.; Murray, R.W.; Hill, C.; Carolina, N. Gateway Reactions to Diverse, Polyfunctional Monolayer-Protected Gold Clusters. *J. Am. Soc.* **1998**, *120*, 4845–4849. [CrossRef]
6. West, J.L.; Halas, N.J. Applications of nanotechnology to biotechnology: Commentary. *Curr. Opin. Biotechnol.* **2000**, *11*, 215–217. [CrossRef]
7. Daniel, M.-C.; Astruc, D. Gold Nanoparticles: Assembly, Supramolecular Chemistry, Quantum-Size-Related Properties, and Applications toward Biology, Catalysis, and Nanotechnology. *Chem. Rev.* **2004**, *104*, 293–346. [CrossRef]
8. Link, S.; Mohamed, M.B.; El-Sayed, M. Simulation of the Optical Absorption Spectra of Gold Nanorods as a Function of Their Aspect Ratio and the Effect of the Medium Dielectric Constant. *J. Phys. Chem. B* **1999**, *103*, 3073–3077. [CrossRef]
9. Thioune, N.; Lidgi-Guigui, N.; Cottat, M.; Gabudean, A.-M.; Focsan, M.; Benoist, H.-M.; Astilean, S.; Lamy de la Chapelle, M. Study of gold nanorods—Protein interaction by localized surface plasmon resonance spectroscopy. *Gold Bull.* **2013**, *46*, 275–281. [CrossRef]
10. Yeh, Y.; Ceran, B.; Rotello, V.M. Gold nanoparticles: Preparation, properties, and applications in bionanotechnology. *Nanoscale* **2012**, *4*, 1871–1880. [CrossRef]
11. Amendola, V.; Pilot, R.; Frasconi, M.; Maragò, O.M.; Iatì, M.A. Surface plasmon resonance in gold nanoparticles: A review. *J. Phys. Consens. Matter* **2017**, *29*, 203002. [CrossRef]
12. Shima, K.; Hirota, M.; Fukuda, M.; Tanaka, A. Determination of urinary lysozyme for potential detection of tubular dysfunction in diabetic nephropathy. *Clin. Chem.* **1986**, *32*, 1818–1822. [CrossRef]
13. Levinson, S.S.; Elin, R.J.; Yam, L. Light Chain Proteinuria and Lysozymuria in a Patient with Acute Monocytic Leukemia. *Clin Chem.* **2002**, *48*, 1131–1132. [CrossRef]
14. Harrison, J.F.; Lunt, G.S.; Scott, P.; Blainey, J.D. Urinary Lysozyme, Ribonuclease, and Low-Molecular-Weight Protein in Renal Disease. *Lancet* **1968**, *291*, 371–375. [CrossRef]
15. Klockars, M.; Reitamo, S.; Weber, T.; Kerttula, Y. Cerebrospinal Fluid Lysozyme in Bacterial and Viral Meningitis. *Acta Med. Scand.* **1978**, *203*, 71–74. [CrossRef]
16. Sreerama, N.; Woody, R.W. Estimation of protein secondary structure from CD spectra: Comparison of CONTIN, SELCON and CDSSTR methods with an expanded reference set. *Anal. Biochem.* **2000**, *287*, 252–260. [CrossRef]
17. Wojdyr, M. Fityk: A general-purpose peak fitting program. *J. Appl. Crystallogr.* **2010**, *43*, 1126–1128. [CrossRef]
18. Yang, S.; Shian, M.S.; Chang, F.Y. Evolution study of photo-synthesized gold nanoparticles by spectral deconvolution model: A quantitative approach. *New J. Chem.* **2006**, *30*, 729–735. [CrossRef]
19. Chen, Y.; Yu, C.; Cheng, T.; Tseng, W. Colorimetric Detection of Lysozyme Based on Electrostatic Interaction with Human Serum Albumin-Modified Gold Nanoparticles. *Langmuir* **2008**, *24*, 3654–3660. [CrossRef]
20. Kim, T.; Blanch, H.W.; Radke, C.J. Direct Imaging of Lysozyme Adsorption onto Mica by Atomic Force Microscopy. *Langmuir* **2002**, *18*, 5841–5850. [CrossRef]

21. Da Paoli Lacerda, H.; Park, J.J.; Meuse, C.; Pristinski, D.; Becker, M.L.; Karim, A.; Douglas, J.F. Interaction of Gold Nanoparticles with Common Human Blood Proteins. *ACS Nano* **2010**, *4*, 365–379. [CrossRef] [PubMed]
22. Lundqvist, M.; Stigler, J.; Cedervall, T.; Berggard, T.; Flanagan, M.B.; Lynch, I.; Elia, G.; Dawson, K. The Evolution of the Protein Corona around Nanoparticles: A Test Study. *ACS Nano* **2011**, *5*, 7503–7509. [CrossRef]
23. Kelly, S.M.; Jess, T.J.; Price, N.C. How to study proteins by circular dichroism. *Biochim. Biophys. Acta* **2005**, *1751*, 119–139. [CrossRef] [PubMed]
24. Benesi, H.A.; Hildebrand, J.H. A Spectrophotometric Investigation of the Interaction of Iodine with Aromatic Hydrocarbons. *J. Am. Chem. Soc.* **1949**, *71*, 2703–2707. [CrossRef]
25. Liu, X.; Atwater, M.; Wang, J.; Huo, Q. Extinction coefficient of gold nanoparticles with different sizes and different capping ligands. *Colloids Surf. B Biointerfaces* **2007**, *58*, 3–7. [CrossRef]

MDPI
St. Alban-Anlage 66
4052 Basel
Switzerland
Tel. +41 61 683 77 34
Fax +41 61 302 89 18
www.mdpi.com

Nanomaterials Editorial Office
E-mail: nanomaterials@mdpi.com
www.mdpi.com/journal/nanomaterials